Clinical Pharmacist's Guide to Biostatistics and Literature Evaluation

Clinical Pharmacist's Guide to Biostatistics and Literature Evaluation

Robert DiCenzo, Pharm.D., FCCP, BCPS

American College of Clinical Pharmacy

Lenexa, Kansas

Director of Professional Development: Nancy M. Perrin, M.A., CAE
Associate Director of Professional Development: Wafa Dahdal, Pharm.D., BCPS (AQ Cardiology)
Publications Project Manager: Janel Mosley
Desktop Publisher/Graphic Designer: Jen DeYoe, B.F.A.
Medical Editor: Kimma Sheldon, Ph.D., M.A.

Material appearing in this book is adapted from matter originally published in the following chapters:

DeYoung GR. Biostatistics: a refresher. In: Bressler L, DeYoung GR, El-Ibiary S, et al., eds. Updates in Therapeutics: The Pharmacotherapy Preparatory Course. Lenexa, Kans.: American College of Clinical Pharmacy, 2010:133–58.

DeYoung GR. Clinical trials: fundamentals of design and interpretation. In: Bressler L, DeYoung GR, El-Ibiary S, et al., eds. Updates in Therapeutics: The Pharmacotherapy Preparatory Course. Lenexa, Kans.: American College of Clinical Pharmacy, 2010:159–80.

Strassels SA. Biostatistics. In: Dunsworth T, Richardson M, Chant C, et al., eds. Pharmacotherapy Self-Assessment Program, 6th ed. Science and Practice of Pharmacotherapy I. Lenexa, Kans.: American College of Clinical Pharmacy, 2007:1–16.

Tsuyuki RT, Garg S. Interpreting data in cardiovascular disease clinical trials: a biostatistical toolbox. In: Richardson M, Chant C, Chen JWM, et al., eds. Pharmacotherapy Self-Assessment Program, 7th ed. Cardiology. Lenexa, Kans.: American College of Clinical Pharmacy, 2010:241–55.

Strassels SA, Wilson JP. Pharmacoepidemiology. In: Dunsworth T, Richardson M, Chant C, et al., eds. Pharmacotherapy Self-Assessment Program, 6th ed. Science and Practice of Pharmacotherapy I. Lenexa, Kans.: American College of Clinical Pharmacy, 2007:17–31.

For order information or questions, contact:
American College of Clinical Pharmacy
13000 West 87th Street Parkway, Suite 100
Lenexa, Kansas 66215
(913) 492-3311
(913-492-4922 (Fax)
accp@accp.com

Printed in the United States of America
ISBN: 978-1-932658-78-1
Library of Congress Control Number: 2011921758

Contents

Clinical Pharmacist's Guide to Biostatistics and Literature Evaluation

HELPFUL DEFINITIONS AND EQUATIONS

Measure	Definitions and Equations
Absolute risk (AR)	Incidence for exposed subjects – incidence for unexposed subjects
Alpha (α)	Probability of a type I error
Attributable fraction	((RR – 1)/RR) × 100%
Beta (β)	Probability of a type II error
Coefficient of determination (r^2)	Proportion of dependent variable variance associated with changes in the independent variable
Confidence interval (CI)	1.96 × SEM
Correlation (r)	Strength and direction of a general linear relationship between two variables
Incidence rate	Number of incident cases/amount of at-risk experience
Interquartile range (IQR) or middle 50% of data	Interval between the 75th and 25th percentile
Mean	Sum of all scores/number of scores
Median	Data midpoint (50% of values above and 50% of values below)
Mode	Most commonly occurring value
Number needed to treat (NNT)	1/AR (expressed as a decimal)
Odds ratio (OR)	*ad/bc* (from a 2 × 2 contingency table)
p value	Likelihood of observing the results by chance alone; or, likelihood of a type I error
Power	1 – β
Prevalence	Number of prevalent cases/size of population
Range	Interval between minimal and maximal value
Relative risk (RR) or risk ratio	Incidence for exposed subjects/incidence for unexposed subjects
Relative risk reduction (RRR)	(Percentage with $\text{event}_{\text{Control Group}}$ – percentage with $\text{event}_{\text{Treatment Group}}$)/percentage with $\text{event}_{\text{Control Group}}$
Risk	Number of subjects developing disease or event during period/number of subjects observed for the period
Risk difference	Incidence for exposed subjects – incidence for unexposed subjects
Standard deviation (SD)	Square root of the variance
Standard error of the mean (SEM)	$\frac{\text{SD}}{\sqrt{n}}$
Type I error	Null hypothesis rejected when it is true
Type II error	Failure to reject the null hypothesis when a difference exists in the sampled population
Variance	Variation in data set for one variable

BASICS OF BIOSTATISTICS AND STATISTICAL TESTS

Basics of Biostatistics

INTRODUCTION

Whether you are interpreting the medical literature to optimize patient care, improve health outcomes, or generate hypothesis for research, an understanding of biostatistics is essential for success. Despite their exposure to biostatistics in undergraduate and professional education, pharmacists tend to be less confident in their knowledge of biostatistics and their ability to interpret the medical literature than in their clinical skills. This review is intended to support pharmacists' preparation for the Pharmacotherapy and Ambulatory Care Board of Pharmacy Specialties examinations.

RANDOM VARIABLES

Continuous Variables

When designing a study, we must first identify the outcomes we wish to measure. Outcome variables should be random, and they will influence the type of statistical test used for analysis. For continuous variables, an infinite number of values are possible within a specific range. Continuous variables can be either interval or ratio. Interval data are ranked in a specific order for which the change in size between units is consistent; ratio data are similar to interval data but have an absolute zero. For example, heart rate has an absolute zero because a heart rate of zero means no pulse rate exists.

Discrete Variables

Discrete variables are countable. They can be dichotomous (one of two possible states) or categorical. Nominal variables are discrete variables classified into groups with no particular order, whereas ordinal variables allow rank in a specific order but without a consistent size of difference between categories. Examples of nominal data include sex and mortality, whereas an example of ordinal data is the trauma score.

DESCRIPTIVE STATISTICS

Central Tendencies

Descriptive estimates provide a general view of data and a means to summarize results. The mean, median, and mode each provide different information about the center of the data and are referred to as measures of central tendency; however, the mean is sensitive to outliers, and comparing the mean with the median can tell the reader about the distribution of the data. For example, if the mean is greater than the median, the data may be skewed to the right. Although we commonly refer to the arithmetic mean, there are variations on this theme. An example is the geometric mean, the average of the logarithmic values of a data set. In regression analyses, the dependent variable may be transformed to its logarithm to improve the way in which the model satisfies underlying assumptions. Using logarithmic transforms is common when reporting or comparing pharmacokinetic parameters such as the area under the concentration-time profile curve (AUC). Like many biologic parameters, the AUC is commonly skewed to the right, yet the log transforms are normally distributed. This approach is common in bioequivalence and drug interaction trials because using logarithmic transforms of the AUC allows the use of parametric tests, which results in increased power and decreased sample size. The final results are reported as the antilog to make them understandable. For example, the logarithmic mean of the AUC is reported as the antilog or geometric mean.

Data Spread or Distribution

Distributions generally describe the spread of the variables. The normal distribution is well known to clinicians, but many types of data encountered in clinical practice are not normally distributed, such as the AUC described above. Measures of variability, including the interquartile range (IQR), range, standard deviation (SD), and variance, are useful in summarizing the spread of the data. The range is the interval between the minimum and maximum value, whereas the IQR is the interval between the 75th percentile and 25th percentile values. The IQR describes the middle 50% of the data in the sample. The variance is the measure of variation in a data set for one variable, and the SD is the square root of the variance. The mean plus or minus 2 SDs will include the central 95% of values.

Although commonly confused with the SD, the standard error of the mean (SEM) is an estimate of certainty that a calculated sample mean represents the true mean of the population. As such, the SEM is an inferential statistic and is not interchangeable with the SD, even though it is sometimes used in this way because it gives the appearance of a more certain knowledge of the central tendencies. Because the SEM is calculated by dividing the SD by the square root of the number of individuals in a data set $\left(\frac{\text{SD}}{\sqrt{n}}\right)$, the SEM will always be smaller than the SD, and it can be erroneously interpreted as indicating that a set of observations is less variable than it really is. When to apply the SD and when to apply the SEM is quite simple. If an investigator is describing individual observations, such as in the patient characteristics table, the SD is appropriate because it is the correct measure of variability of individual observations; however, if an investigator is reporting the mean of a study outcome, the SEM is the correct measure of variability because it is the variability, or uncertainty, in the estimate of the mean.

Discrete Distributions

The binomial and Poisson distributions are two discrete probability distributions. The binomial distribution is used when considering a sample of some number of independent trials that have only two possible outcomes, such as some indeterminate measure of success or failure. Imagine that a coin is flipped 1,000 times. Each time the coin is flipped, there is some probability of success or failure. The Poisson distribution is used most commonly when rare events are being considered. An example is the number of serious adverse drug reactions caused by drug A during some period. Approximations to the binomial or Poisson distribution are used when use of the distribution under question would be onerous and when specific conditions exist. For example, the binomial distribution is useful when considering some number of independent trials. But if the number of trials performed is large and the probability of success is either very high or very low, the distribution will be skewed. If, however, the number of trials undertaken is at least moderately large and the probability of success is not too extreme, the distribution will be symmetric and will be approximated by the normal distribution. Similarly, when the expected number of events during a time interval of interest is large, the Poisson distribution is unwieldy, and a similar normal approximation may apply.

Continuous Distributions

The best-known example of a continuous distribution is the normal distribution. Most clinicians are familiar with at least some features of this distribution, which is also referred to as a gaussian distribution or a bell-shaped curve. It is symmetric around the mean, and when the mean is zero with a variance (and, thus, an SD as well) of 1, it is referred to as a standard normal distribution. The area under the standard normal curve from 1 SD below the mean to 1 SD above the mean includes about 68% of the distribution, whereas 95% of the distribution lies in the area from 2 SD below the mean to 2 SD above, and 99% of the area is from 2.5 SD below the mean to 2.5 SD above the mean (Figure 1).

Central Limit Theorem

The central limit theorem states that in considering a random sample, when the number of observations is large, the distribution of the mean is about normally distributed, even if the distribution of the observations in the sample being studied is not normally distributed. Often, 30 observations are used as a rule-of-thumb minimal cutoff for defining what is meant by "large." Normally distributed data are rare in real life, but if there is reason to believe that the central limit theorem applies, inferential statistics can be used. In practice, this is important because the means

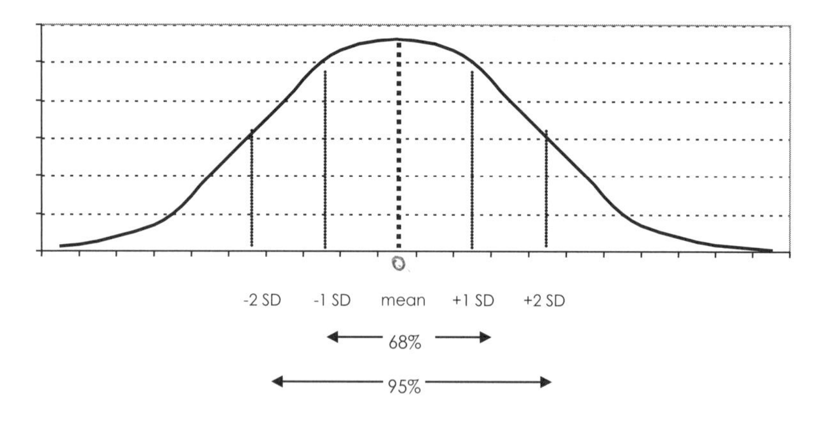

Figure 1. Standard deviation.

SD = standard deviation.

of most physiologic measurements, such as blood pressure, are considered to be normally distributed.

Degrees of Freedom

Many statistical tests depend on a factor called the degrees of freedom (*df*). The *df* is the number of individual observations that are "free to vary." Calculations that use the *df* include the variance and SD. For example, the SD is the mean distance of an observation from the sample mean. Because the sample mean is included in the formula, one observation is no longer free to vary; therefore, it should not be counted in the number of observations that are free to contribute to the variability.

HYPOTHESIS TESTING

Hypothesis testing involves creating a null hypothesis (H_0) and an alternative hypothesis (H_A) and then calculating a value (statistic) from the observed data. If a difference exists between the study groups, the H_0 is rejected. This occurs when the calculated statistic does not appear to belong to a distribution (range of values) for the test statistic when the H_0 is true. If no difference between groups is found, the result is a failure to reject the H_0 versus accepting the H_A because every study is limited, and it is impossible to know whether there are truly no differences in the factor of interest between groups. Testing for equivalence involves completely different hypothesis testing and study design. The H_0 and H_A are typically expressed by saying that the groups are equal (or that there is no difference between them) and that the groups are not equal (or that there are some differences between groups), respectively, though more specific ways of expressing those ideas are also used. For example, the H_A is often stated as a general inequality (i.e., the mean of group 1 is not equal to the mean of group 2) because the direction of the inequality is uncertain. In the uncommon case in which certainty exists about the direction of differences between groups, the researcher may choose to express the H_A in that direction (e.g., the mean of group 1 greater than the mean of group 2).

The potential for making errors when evaluating the H_0 and H_A using inferential statistics is well known (Table 1). A type I error occurs when the H_0 is rejected when it is true, thus finding differences where none exists. A type II error is a failure to reject the H_0, even though a difference is

Table 1. Hypothesis Testing

Your Decision	Underlying "Truth"	
	H_0 Is True	*H_0 Is False*
Accept H_0	No error	Type II error
Reject H_0	Type I error	No error

H_0 = null hypothesis.

present. The probability of a type I error, or the significance level of a test, is denoted by alpha (α), and the probability of a type II error is referred to as beta (β).

β is a component of the power of a test, which contributes to determining the sample size needed to detect a difference between groups if one exists. Power is calculated as 1 − β. The importance of the estimated power of a study is highlighted when the observed difference between groups is not statistically significant, because the failure to find a difference could be owing to similarity between groups or insufficient power. The power of a test is affected by several factors including the significance level (α), the difference between groups, the variance of the observations, and the sample size. Power increases as the difference between groups and the sample size increases. Power decreases as α falls, and the variance of the observations increases.

Sample size is another important consideration when designing a trial or interpreting the literature. Sample size depends on many factors, including the significance level, variance, desired power, and size of difference determined to be clinically meaningful. The number of study participants needed to detect a given difference increases as the variance in the data increases, the chosen α decreases, and the power increases (or β decreases). For example, suppose that researchers want to conduct a study in which they are willing to accept a 1% probability that observed differences between groups are solely attributable to chance. If the investigators are willing to accept a 5% probability, instead of just 1%, that differences are attributable to chance, the needed sample size will decrease. The implications of these types of decisions are important because a larger sample increases the time and expense needed to conduct a clinical study.

By convention, biomedical researchers often set $\alpha = 0.05$ and $\beta = 0.20$, though there is no reason that different values of each cannot be used. For example, because making an error can have important consequences, there may be compelling reasons to avoid making a type I error, even if that means making a type II error, or vice versa. Imagine that researchers develop a new screening test to detect colorectal cancer. In a clinical trial, the investigators find that the new test is more sensitive and specific than the existing test, despite being more invasive and expensive. If the investigators have made a type I error and there is really no difference between the tests, the consequences include exposing a person to the unneeded risks of adverse effects because of the invasiveness of the test. Alternately, if a type II error is made and the tests are found to be equivalent when they are not, patients may be unable to take advantage of the improved technology.

One- and Two-Tailed Tests

A one-tailed statistical test refers to one in which the parameter being studied (such as the mean of some variable) under the H_A is allowed to be either less than or greater than the values under the H_0, but not both. By contrast, in a two-tailed test, the values of the parameter of interest under the H_A are allowed to be less than or greater than the values under the H_0.

Deciding which approach to use requires some thought before the data are analyzed. It is acceptable to test for differences in either direction, in which case the H_0 will be rejected if the value of the test statistic is above or below the critical point. This is tantamount to saying that researchers will conclude that an observed difference is statistically significant if the mean value for group 1 is above or below that for group 2. In general, however, using a one-tailed test is analogous to stating that there is no interest in changes in the other direction. In addition, justifying the decision to use a one-tailed approach often requires information that is unavailable. Although the two-tailed approach is more common in the literature, the one-tailed approach is valid if justified.

P Values and Confidence Intervals

The calculated chance that a type I error has occurred is called the p value. A p value is the probability of obtaining a result at least as extreme as the one observed, given that the H_0 is true. Reporting a p value of 0.05

effectively means that 1 in 20 times, a type I error will occur when the H_0 is rejected. This definition is problematic, however. P values are conditional on the H_0, but it is unknown whether the H_0 is actually true. Furthermore, p values are calculated with models that correspond to the type of data used, and most models assume that observations are independent. Yet, depending on the type of study, this assumption may not be true. As a result, a p value is generally not a meaningful estimate of probability, but more an indication of consistency between the H_0 and the data. The result is that a large p value suggests the data are consistent with the H_0, and a small p value suggests the data are inconsistent with the H_0. Neither, however, tells us whether the H_0 is true.

There are many ways to interpret p values. Assuming that the significance level for a study is 0.05, one approach is to view p values between 0.01 and 0.05 as significant, those between 0.001 and 0.01 as highly significant, those less than 0.001 as very highly significant, and those of 0.05 or higher as not statistically significant. The p values higher than 0.05 but lower than 0.10 can be referred to as trending toward significance, but trends represent value judgments. A trend toward significance can also easily be interpreted as a trend away from significance. Because the significance level is arbitrary, statistical significance tells only that the p value is less than the cutoff value chosen. Furthermore, significance testing, with its dichotomous outcome, provides no information about the size of the effect or its clinical significance. Similarly, a statistically insignificant p value does not indicate that there is no association in the data.

Together with p values, confidence intervals (CIs) are used to account for random error. As with significance levels, the desired level of confidence is chosen arbitrarily. A 95% CI, often used by convention in biomedical research, indicates that, if the experiment were to be repeated many times, at least 95% of the resulting CIs constructed would include the true (unobservable) population mean. But this interpretation assumes that the statistical model being used is correct and that bias is negligible. These assumptions may not hold in all types of medical research. As a result, the CI should be considered only a general and minimal estimate of uncertainty in the estimate. In addition, the higher the confidence level, the wider the interval will be, because we need to include a wider range of values to be more certain that the true value of the variable of interest is included. Confidence intervals can be calculated in several ways. One means of estimating the limits of a 95% CI is by multiplying 1.96 by the SEM and then adding and subtracting that quantity to the mean. The factor, 1.96, comes from the probability that a value will fall within about 2 SDs to either side of the mean under a normal distribution.

The connection between p values and confidence levels is obvious, but it represents an important pitfall. Although CIs can be viewed as analogous to hypothesis tests of significance (e.g., does the CI of an odds ratio [OR] include 1 or the CI of a difference between groups include zero), doing so is pointless and ignores the information CIs offer above what hypothesis testing provides. For example, the CI estimates the effect size as well as the variability in the estimate, whereas the p value provides only an estimate of the consistency between the data and the hypothesis. Therefore, CI estimates are more informative because they report significance and provide an estimate of the size of the effect, allowing the reader to better apply the results to his/her clinical practice.

Incidence Rates, Prevalence Rates, Odds, and ORs

Incidence rates, prevalence rates, odds, and ORs are commonly encountered in epidemiologic research. The incidence rate is an estimate of the instantaneous rate of developing disease. It is calculated by dividing the number of individuals who develop disease in a population during a given time by the summed amount of time that people in the study were at risk of developing the disease,

$$\frac{\text{number of subjects developing disease}}{\text{total time at risk of disease for subjects observed}}$$

If 100 people are observed for a year and are at risk of disease for that time, the denominator of the incidence rate will be 100 person-years.

Incidence reflects the rate at which people develop disease, whereas prevalence estimates the number of people who have a condition at a particular time. Like incidence, prevalence is an important epidemiologic measure. In addition, by dividing the proportion of individuals with a disease by the proportion of individuals without the disease $\left(\frac{P}{1-P}\right)$, it is possible to estimate the prevalence odds. This relation holds for other measures as well. In a case-control study, the OR is the ratio of cases to controls among the exposed individuals, divided by the ratio of cases to controls among the unexposed individuals. An OR of 1.0 indicates that the independent variable is not associated with the outcome. Values greater than 1.0 indicate that the independent variable is associated with a higher risk of the outcome, whereas values lower than 1.0 indicate that exposure to the independent variable is associated with a lower risk of the outcome. In case-control studies, control subjects can be selected using different methods. In one such method, called incidence-density, or risk-set, sampling, control subjects are selected from among all individuals at risk of the event when the event occurred. When this method is used, the control subjects are chosen from among individuals at risk of experiencing the outcome when the event of interest occurs for an individual. Under this condition, the OR provides a valid estimate of the incidence rate ratio.

Contingency Tables

Contingency tables are useful to estimate the association between variables. When constructed to examine the relationship between two variables, the table includes two rows and two columns, though any number of rows or columns can be used to accommodate variables with more than two categories (Table 2). Within this 2 × 2 table, the variables must be categorized so that each term has only two possible results. One variable is arbitrarily assigned to the rows of the table, and the other is assigned to the columns. Each cell of the table contains the number of individuals who meet the criteria for both variables, such as exposed and with outcome or not exposed and without outcome. By convention, row and column totals are calculated and written in the right and bottom margins, respectively. The grand total, or the sum of all individuals in the table, is also written in the lower right-hand corner of the table.

The odds ratio is estimated by calculating $\frac{AD}{BC}$. The expected value for a particular cell is calculated as the product of the row total and the column total, divided by the grand total. Using the table above, the expected

Table 2. 2 × 2 Contingency Table

Exposure	Outcome: yes	Outcome: no	
Yes	A	B	Row 1 total (R1)
No	C	D	Row 2 total (R2)
	Column 1 total (C1)	Column 2 total (C2)	Grand Total (GT)

value for cell A is $\frac{R_1C_1}{\text{GT}}$. Expected values for the other cells are calculated the same way.

Before performing statistical tests using a contingency table of any size, the expected values for each of the cells are calculated, which represent expected counts if the H_0 is true. This information permits comparisons of expected and observed cell totals, which provides an opportunity to visually evaluate the closeness of the two types of data. The expected value for each cell is the product of the row and column totals divided by the grand total. In addition, two other measures can be estimated from observed and expected values. First, by cross-multiplying and dividing the observed estimates $\frac{\text{AD}}{\text{BC}}$, we can estimate the OR. In a case-control study in which risk-set sampling has been used, the OR is an estimate of the incidence rate ratio. Second, the standardized mortality (or morbidity) ratio can be estimated by dividing the observed number of events by the expected number of events and multiplying that number by 100%. Values of this ratio that are less than, equal to, or greater than 100% indicate that the risk of the outcome in the study population is reduced, the same as, or larger than that of the general population, respectively.

SELECTED BIBLIOGRAPHY

Altman DG, Bland JM. Interaction revisited: the difference between two estimates. BMJ 2003;326:219.

Altman DG, Bland M. Standard deviations and standard errors. BMJ 2005;331:903.

Bradburn MJ, Clark TG, Love SB, Altman DG. Survival analysis part II: multivariate data analysis—an introduction to concepts and methods. Br J Cancer 2003;89:431–6.

Bradburn MJ, Clark TG, Love SB, Altman DG. Survival analysis part III: multivariate data analysis—choosing a model and assessing its adequacy and fit. Br J Cancer 2003;89:605–11.

Braitman LE. Confidence intervals assess both clinical significance and statistical significance. Ann Intern Med 1991;114:515–7.

Centre for Health Evidence and JAMA and Archives. Users' Guides to the Medical Literature. Available at http://pubs.ama-assn.org/misc/usersguides.dtl. Accessed August 21, 2007.

Clark TG, Bradburn MJ, Love SB, Altman DG. Survival analysis part I: basic concepts and first analyses. Br J Cancer 2003;89:232–8.

Clark TG, Bradburn MJ, Love SB, Altman DG. Survival analysis part IV: further concepts and methods in survival analysis. Br J Cancer 2003;89:781–6.

DeYoung GR. Understanding biostatistics: an approach for the clinician. In: Zarowitz B, Shumock G, Dunsworth T, et al., eds. Pharmacotherapy Self-Assessment Program. The Science and Practice of Pharmacotherapy I Module, 5th ed. Kansas City, MO: ACCP, 2005:1–20.

Gaddis ML, Gaddis GM. Introduction to biostatistics. Part 1, basic concepts. Ann Emerg Med 1990;19:86–9.

Gaddis ML, Gaddis GM. Introduction to biostatistics. Part 2, descriptive statistics. Ann Emerg Med 1990;19:309–15.

Gaddis ML, Gaddis GM. Introduction to biostatistics. Part 3, sensitivity, specificity, predictive value, and hypothesis testing. Ann Emerg Med 1990;19:591–7.

Gaddis ML, Gaddis GM. Introduction to biostatistics. Part 4, statistical inference techniques in hypothesis testing. Ann Emerg Med 1990;19:820–5.

Gaddis ML, Gaddis GM. Introduction to biostatistics. Part 5, statistical inference techniques for hypothesis testing with nonparametric data. Ann Emerg Med 1990;19:1054–9.

Gaddis ML, Gaddis GM. Introduction to biostatistics. Part 6, correlation and regression. Ann Emerg Med 1990;19:1462–8.

Glantz SA. Primer of Biostatistics, 5th ed. New York: McGraw-Hill, 2002.

Hewitt CE, Mitchell N, Torgerson DJ. Heed the data when results are not significant. BMJ 2008;326:23–5.

Hosmer DW Jr, Lemeshow S. Applied Survival Analysis: Regression Modeling of Time to Event Data. New York: John Wiley & Sons, 1999.

Hosmer DW Jr, Lemeshow S. Applied Logistic Regression, 2nd ed. New York: John Wiley & Sons, 2000.

Kleinbaum DG, Klein M. Survival Analysis: A Self-Learning Text, 2nd ed. New York: Springer, 2005.

Kleinbaum DG, Klein M, Pryor ER. Logistic Regression: A Self-Learning Text, 2nd ed. New York: Springer, 2005.

Kleinbaum DG, Kupper LL, Muller KE, Nizam A. Applied Regression Analysis and Multivariable Methods, 3rd ed. New York: Duxbury Press, 1998.

Levine M, Ensom MHH. Post hoc power analysis: an idea whose time has passed? Pharmacotherapy 2001;21:405–9.

Rosner B. Fundamentals of Biostatistics, 6th ed. New York: Duxbury Press, 2005.

Statistics Series in Critical Care: Articles appear beginning February 2002 through January 2005. Includes the following topics: I. Presenting and summarizing data; II. Samples and populations; III. Hypothesis testing and p-values; IV. Sample size calculations; V. Comparison of means; VI. Nonparametric methods; VII. Correlation and regression; VIII. Qualitative data—tests of association; IX. One-way analysis of variance; X. Further non-parametric methods; XI. Assessing risk; XII. Survival analysis; XIII. Receiver operating characteristic curves; and XIV. Logistic regression. Available at *http://ccforum.com/articles/stats-series.asp*. Accessed September 30, 2009.

SELF-ASSESSMENT QUESTIONS

1. In a study reporting patient satisfaction with care, participants were asked to rate their satisfaction with their medical care on a 0–5 Likert-type scale (0 = not satisfied at all, 5 = completely satisfied). Which one of these measures is best to describe patients' response to this question?

   A. Mean
   B. Interquartile range
   C. Median
   D. Mode

Questions 2 and 3 pertain to the following case.
In a randomized, controlled trial of analgesia after total joint replacement surgery, the median difference between treatment groups in the 0–10 numeric rating scale of pain intensity on postoperative day 2 was 2.2 (95% confidence interval [CI], 1.1–3.5; p=0.01).

2. Which one of the following reasons best summarizes why the p value or CI is preferred?

   A. The p value is preferred because it provides an estimate of the magnitude of the observed difference.
   B. The CI is preferred because it provides an estimate of the magnitude of difference.
   C. The p value is preferred because it provides the insight needed for clinical decision-making.
   D. The CI is preferred because it provides the insight needed for clinical decision-making.

3. How would this point estimate and CI be interpreted?

   A. The first treatment provided better management of postoperative pain than the second treatment.
   B. The second treatment provided better management of postoperative pain than the first treatment.
   C. No statistically significant difference was observed between the two treatments.
   D. One treatment was more statistically significant than the other.

Questions 4 and 5 pertain to the following case.

In a randomized, controlled study of preemptive analgesia in men who underwent radical prostatectomy surgery, functioning was statistically ($\alpha=0.05$) and clinically significantly improved in men who received the experimental treatment compared with men who received usual care.

4. Which one of the following can be said about the assessment of statistical significance in this case?
   A. The p value was less than or equal to 0.05.
   B. The p value is an absolute indicator of which treatment was better.
   C. The assessment of statistical significance is independent of the significance level.
   D. Statistical significance was more important than clinical significance.

5. Which one of the following can be said of clinical significance in this case?
   A. Clinical significance was less important than statistical significance.
   B. The difference between treatments required to be judged clinically significant is likely to be smaller than that needed for statistical significance.
   C. The difference between treatments required to be judged clinically significant is likely to be larger than that needed for statistical significance.
   D. Clinical and statistical significance are equally important.

6. In a regression model, individuals with cancer had an odds ratio (OR) of 1.5 (95% CI, 1.2–2.0) compared with individuals without cancer. Which one of the following is the best way to interpret these results?
   A. Having cancer increased the likelihood that the person experienced the outcome by 150%.
   B. Having cancer increased the likelihood that the person experienced the outcome by 20% to 100%.
   C. Having cancer decreased the likelihood that the person experienced the outcome by 150%.
   D. The estimated OR did not reflect a statistically significant result.

Questions 7 and 8 pertain to the following case.

In a randomized, controlled trial, investigators assessed whether there were overall differences between two drug treatments. Ninety people completed the trial, divided equally between treatments. Of those who received drug A, 29 experienced the outcome, whereas of those who received drug B, 14 experienced the outcome.

7. Which one of the following is the best estimate of the OR for this study comparing the experience of individuals who received drug A with those who received drug B?

    A. 4.0
    B. 0.25
    C. 1.1
    D. 0.9

8. Which one of the following statements is the best interpretation for an OR of 4.0?

    A. The odds of having the outcome among individuals who received treatment A are 10% higher than the odds among individuals who received treatment B.
    B. The odds of having the outcome are 75% lower in individuals who received treatment A than those in individuals who received treatment B.
    C. The odds of having the outcome among individuals who received treatment B are 10% lower than the odds among individuals who received treatment A.
    D. The odds of having the outcome are 400% higher among individuals who received treatment A than those among individuals who received treatment B.

9. In a meta-analysis of randomized, controlled trials examining the effects of antihypertensive drugs, investigators found that the OR for treatment with low-dose diuretics compared with β-blockers for cardiovascular disease events was 0.86 (95% CI, 0.77–0.97). Which one of the following statements is the most appropriate interpretation of these findings?

    A. Treatment of hypertension with low-dose diuretics was 14% more effective in preventing cardiovascular disease events than treatment with β-blockers.

B. Treatment of hypertension with β-blockers was 14% more effective in preventing cardiovascular disease events than treatment with low doses of diuretics.
C. The difference observed between treatment with β-blockers and low doses of diuretics was not statistically significant.
D. Treatment of hypertension with low doses of diuretics was from 3% to 23% more effective in preventing cardiovascular events than treatment with β-blockers.

Questions 10–13 pertain to the following case.

In preparation for constructing a research poster for a national pharmacy meeting, you and a resident are analyzing data from a pilot project of systolic blood pressure (SBP) and oral contraceptive use in 16 women. The SBP readings for each person before and after starting oral contraceptive use are shown in the table below. Assume that SBP is normally distributed at baseline.

Patient No.	SBP While Not Using Oral Contraceptives (mm Hg)	SBP While Using Oral Contraceptives (mm Hg)
1	120	133
2	117	120
3	112	111
4	124	133
5	120	127
6	143	150
7	131	137
8	110	114
9	109	107
10	124	126
11	110	123
12	99	102
13	100	99
14	121	130
15	133	140
16	109	116

SBP = systolic blood pressure.

10. Which one of the following choices is the best interpretation of the mean difference in SBP in study participants?
    A. Although study participants used oral contraceptives, about 50% of them had an increase in SBP of 6.5 mm Hg.
    B. Although study participants used oral contraceptives, their SBP increased by 5.4 mm Hg.
    C. Although study participants used oral contraceptives, SBP for most of them increased by 7.0 mm Hg.
    D. Although study participants used oral contraceptives, their SBP readings increased between 5.4 mm Hg and 7.5 mm Hg.

11. The standard deviation (SD) of the mean of the pairwise differences, on treatment SBP minus before treatment SBP for each patient, is 4.6. Which one of the following choices is the best interpretation of the standard error of the mean (SEM) difference in SBP?
    A. The average distance of each value in this data set from its expected value is 5.29.
    B. The average distance of each value in this data set from its expected value is 1.32.
    C. In general, the sample mean differs from a set of sample means by 0.29.
    D. In general, the sample mean differs from a set of sample means by 1.15.

12. Which one of the following statistical tests would be most appropriate for analyzing this data set?
    A. Chi-square
    B. t-test
    C. Paired t-test
    D. Cox proportional hazards model

13. Assume that in a related study assessing this question in 25 women, the mean difference was 10 mm Hg, the median difference 6 mm Hg, and the SEM difference 2.0. Which one of the following choices is the best interpretation of the 95% CI for the mean difference?

A. If the data collection and analysis were repeated many times, the interval between 8.0 and 12.0 would contain the mean difference in 95% of those trials.
B. If the data collection and analysis were repeated many times, the interval between 6.1 and 13.9 would contain the mean difference in 95% of those trials.
C. If the data collection and analysis were repeated many times, the interval between 4.0 and 8.0 would contain the mean difference in 95% of those trials.
D. If the data collection and analysis were repeated many times, the interval between 2.1 and 9.9 would contain the mean difference in 95% of those trials.

14. You wish to describe the type of patients who use the anticoagulation monitoring services your department provides. Specifically, the Pharmacy and Therapeutics Committee is interested in the age of the population served. Which one of the following pieces of information will be most useful to the committee members?
    A. Standard error of the mean (SEM)
    B. t-test
    C. Mean
    D. Chi-square test

As part of a process improvement committee, you are responsible for determining the impact of a recent educational campaign to improve the recording of patient allergies in the medical record. Before the educational efforts, you record the allergy status of 100 patients on one unit using their admission orders. After the education, you assess the allergy status of another 100 patients from another unit using the same method. Your results are as follows:

	Allergy Recorded	Allergy Not Recorded
Before education	78	22
After education	90	10

15. In preparing to analyze the results of this intervention statistically, you consider the data in the table which one of the following types?

    A. Nominal
    B. Ordinal
    C. Interval
    D. Ratio

16. In a cohort study designed to determine an association between measles, mumps, and rubella vaccination and autism, investigators report the relative risk (RR) of autistic disorder in the vaccinated group compared with the unvaccinated group as 0.92 (95% confidence interval [CI], 0.65–1.07). Which one of the following p values is consistent with these reported findings?

    A. Less than 0.05
    B. Less than 0.01
    C. Greater than 0.05
    D. Greater than 0.10

17. A trial (n=48) reports that the average dose of an intravenous analgesic drug needed to keep postoperative pain below a rating of 2 on a 10-point scale is 67 mg with a standard deviation (SD) of 17 mg. You wish to calculate a CI for this mean dose. Which one of the following is the SEM associated with this result?

    A. 0.35
    B. 1.39
    C. 2.45
    D. 6.93

18. A trial compares drug X and drug Y for treating nausea and vomiting associated with pregnancy because clinicians believe they may differ in their efficacy in preventing nausea and vomiting in pregnant women. Drug X has been used for many years and has a large evidence base showing efficacy and safety. Drug Y was recently introduced to treat nausea and vomiting associated with chemotherapy, but it has not been well studied in patients with nausea and vomiting caused by pregnancy. Patients will be randomized to one of these two drugs.

Which one of the following represents the correct statement of the null hypothesis (H_0) for this trial?

A. The efficacy of drug X equals the efficacy of drug Y.
B. The efficacy of drug X does not equal the efficacy of drug Y.
C. The efficacy of drug X is greater than the efficacy of drug Y.
D. The efficacy of drug X is less than the efficacy of drug Y.

19. A prospective, randomized, double-blind study (n=2200) finds that when two oral drugs for treating type 2 diabetes mellitus are compared, the final mean percent hemoglobin A1C value is 8.21 for patients in group 1 and 8.27 for patients in group 2. This difference between the groups is stated as having a calculated p value less than 0.05. Assuming similar baseline characteristics and appropriate final statistical analysis, which one of the following statements best characterizes these findings?

    A. The difference between the drugs is clinically but not statistically significant.
    B. The difference between the drugs is both statistically and clinically significant.
    C. The difference between the drugs is statistically, but not clinically, significant.
    D. The difference between the drugs is neither statistically nor clinically significant.

20. A case-control study is performed to judge whether a drug is associated with an increased incidence of early miscarriage. The final analysis shows that the OR for miscarriage with drug exposure is 1.3 (95% CI, 0.9–1.7). Which one of the following provides a correct description of these results?

    A. This drug increases the risk of miscarriage by 70%.
    B. This drug increases the risk of miscarriage by 30%.
    C. This drug decreases the risk of miscarriage by 10%.
    D. This drug is not associated with an increased risk of miscarriage.

21. The manager of an obesity clinic in your health care system approaches you about selling a recently marketed herbal weight-loss supplement in the clinic. She tells you that unlike other products

making claims about weight loss, this product has been described to her as containing no ephedra (ma huang) or other stimulants and no dangerous herbal derivatives. She then shows you a copy of a trial proving that this supplement works. You review the study that claims this product increases metabolism. The study shows that patients taking the supplement burned an average of 20 calories more during a 700-calorie workout than those not taking the supplement ($p<0.05$). Which one of the following is an appropriate response to the manager on the basis of the information provided?

A. The statistical differences show that the product is worth using.
B. The differences shown do not appear to be clinically significant.
C. The study shows the product is effective regardless of the p value.
D. The study shows neither statistical nor clinical differences.

Choosing the Appropriate Statistical Test

INTRODUCTION

Choosing the appropriate statistical test is complicated by the number of statistical tests available and the possible availability of many tests for a given instance. To decide whether a test is appropriate, the investigator should consider several factors, including the assumptions underlying a specific test, the number of groups being compared, whether the samples are independent or paired, and whether the data are nominal, ordinal, or continuous. Statistical tests are sometimes described as being parametric or nonparametric. These categories refer to whether the data follow a known distribution and whether certain assumptions about the parameters of the distribution can be made. If appropriate for use, parametric tests have more power than their nonparametric counterparts. Continuous data are generally analyzed using parametric tests, such as the t-test, whereas categorical data are often analyzed using nonparametric tests, such as the chi-square test (χ^2). This section presents statistical tests commonly chosen for use in clinical research.

COMMON STATISTICAL TESTS

Parametric Tests

The z-test

The z-test is used to make an inference about the mean of the sampling when the underlying distribution is normal or the central limit theorem applies, and the variance is known. Yet the variance is often unknown, or the analyst may have reason to believe that the sample and population variances are different. In this case, a t-test, rather than the z-test, is used, resulting in the uncommon use of the z-test in the medical literature.

Paired and Unpaired t-tests

The t-test is used when the means of two groups are compared, given that the assumptions required for the appropriate use of the t-test have been met. Unpaired tests apply when subjects are divided into separate groups for comparison. Paired tests, such as the paired t-test, apply when estimates from one sample are compared with estimates from a matched sample. An example of related data is a pre-post design in which study participants act as their own experimental control. This design has advantages over unpaired tests. Unlike an unpaired test, a paired test eliminates variability between subjects, resulting in the enrollment of fewer subjects while achieving similar power. Examples of the unpaired test and the corresponding paired test appear in Table 3.

Analysis of Variance

Analysis of variance (ANOVA) methods permit comparisons of more than two independent groups. For example, to examine the effect of smoking on the severity of pulmonary disease, rather than examining only the two categories of either no exposure to tobacco smoke or any exposure, the study could assess the relationship between the degree of pulmonary disease and never smokers, passive smokers, former smokers, current light smokers, current moderate smokers, and current heavy smokers. When the effect of one variable on the outcome is analyzed, a one-way ANOVA is used.

Some tests, including the ANOVA, require the investigator to assume that variances between samples are equivalent. To test this assumption, we can assess the ratio of sample variances using the F test. This test is used because the ratio of sample variances follows an F distribution with $n_1 - 1$ and $n_2 - 1$ *df*. When the F test is used for this purpose, it is sensitive to departures from normality.

When the ANOVA indicates that a difference exists among the several groups, further analysis must be performed to determine which groups are different from one another; these tests are commonly referred to as post hoc tests. Although t-tests are used to compare different pairs of groups, performing many comparisons increases the chance of finding a statistically significant difference between groups. To address this issue, multiple comparisons procedures are used to ensure that the chance of finding significant differences between all possible groups is held constant. There are many multiple comparisons procedures, such as the Bonferroni adjustment, the Scheffé test, and the honest significant difference methods.

In the Bonferroni adjustment, the initial significance level is divided by the possible number of independent two-group comparisons. For example, if there are 10 groups, there are 45 possible two-group combinations. If α is set a priori at 0.05, 0.05 divided by 45 gives an adjusted α of 0.0011.

Table 3. Commonly Used Statistical Tests

Type of Variable	2 Samples (independent) (Parallel Design)	2 Samples (related) (Crossover or Pre-Post Design)	≥ 3 Samples (independent) (Parallel Design)	≥ 3 Samples (related) (Crossover Design)
Nominal				
No confounders	χ^2 or Fisher exact test	McNemar test	χ^2 (Bonferroni)	Cochran Q (Bonferroni)
1 confounder	Mantel-Haenszel	Rare	χ^2 (Bonferroni)	Rare
≥ 2 confounders	Logistic regression	Rare	Logistic regression	Rare
Ordinal				
No confounders	Wilcoxon rank sum or Mann-Whitney U-test	Wilcoxon signed rank test	Kruskal-Wallis ANOVA (MCP or Bonferroni)	Friedman ANOVA
1 confounder	2-way ANOVA ranks	2-way repeated ANOVA ranks	2-way ANOVA ranks	2-way repeated ANOVA ranks
≥ 2 confounders	ANOVA ranks	Repeated-measures regression	ANCOVA ranks	Repeated-measures regression
Continuous				
No confounders	Student t-test	Paired Student t-test	1-way ANOVA (MCP)	Repeated-measures ANOVA (MCP)
1 confounder	ANCOVA	2-way repeated ANOVA	2-way ANOVA	2-way repeated ANOVA
≥ 2 confounders	ANCOVA	Repeated-measures regression	ANCOVA	Repeated-measures regression

ANCOVA = analysis of covariance; ANOVA = analysis of variance; MCP = multiple comparison procedures.

Rejecting or failing to reject the H_0 for each comparison would be based on the adjusted α. Note that some of the comparisons may not be independent, in which case the Bonferroni adjustment will be conservative.

The one-way ANOVA is used to estimate the effect of one factor on the dependent variable. In this model, we are interested in estimating the mean of all groups considered together, the difference between the mean of a specific group and the overall mean, and the random error between the overall mean plus the group mean and a single observation. This is also called a one-way ANOVA fixed-effects model because the groups being compared have been fixed by the study design. An alternative to the fixed-effects model approach is the random-effects model. In this variation, an assessment is made on the overall differences between groups and the general breakdown of total variation into between-group and within-group components. Random-effects models allow investigators to analyze subgroup effects in the entire data set without dividing the data into subgroups. Like the fixed-effects model, a random-effects one-way ANOVA model is assessed using the F test. If a study includes both random and fixed variables, then a mixed-effects model can be used.

In addition to the one-way model, there are the two-way and multiple ANOVA approaches. The two-way ANOVA is used when the effects of one variable are analyzed while controlling for the effects of the other variable. This model also allows us to examine whether the effects of a variable on the outcome differ by the level of a second variable. This type of assessment is done by including an interaction term—the product of the variables of interest—in the model and interpreting the results of the statistical test for the two variables. For example, the effects of age and sex on some outcome can be assessed using a two-way ANOVA. To do this, the effects of age and sex (also called the main effects) and a third (interaction) term, age × sex, on the outcome are modeled. By generalizing this approach in an n- (or multi-) way ANOVA, the effect of higher-order interaction terms on an outcome can also be estimated, such as a three-term product (e.g., age × sex × race).

Analysis of Covariance

Analysis of covariance (ANCOVA) is another variation on the ANOVA theme. Like a multi-way ANOVA, an ANCOVA models the effects of more than one independent variable on the outcome. However, an ANOVA model compares the effect of categorical groups on the mean outcome, whereas an ANCOVA provides the flexibility to estimate and control for the effect of continuous independent variables, or covariates, on the outcome. Similarly, it is possible to model more than one outcome in an ANOVA or ANCOVA setting with the multivariate ANOVA or multivariate ANCOVA.

Nonparametric Tests

When data are distributed normally or when normal approximations apply, techniques like those discussed above are commonly used. When assumptions of normality cannot be made and the data do not follow a known parametric distribution, or are categorical, nonparametric methods are used to test hypotheses. Examples of nonparametric tests include the Wilcoxon signed rank and rank sum tests, the chi-square (χ^2) test, and the Kruskal-Wallis test.

Wilcoxon Signed Rank and Wilcoxon Rank Sum Test

The Wilcoxon signed rank test is analogous to the paired t-test. The Wilcoxon signed rank test considers the difference between the observation and the H_0, taking into account the direction and relative size of the observed differences. Instead of the precise magnitude of the difference, the relative magnitude is considered, with greater weight, or higher rank, given to the larger differences. A similar circumstance exists when data are collected from two independent samples. If the data measures are continuous, the t-test for independent samples can be used. When the outcome data are ordinal, however, a nonparametric approach, such as the Wilcoxon rank sum test, should be used.

Chi-square and McNemar Test

Chi-square tests (and variations of it) are commonly used to test hypotheses when data are categorical. The investigator is comparing the expected frequency with the observed frequency to determine the likelihood of the difference being attributable to chance. For example, an investigator is comparing the frequency of lung cancer development between smokers and nonsmokers to determine whether the difference in frequency is attributable to chance. For a 2 × 2 cross-tabulation table, all expected values must be at least 5 to use the χ^2 test; otherwise, the Fisher exact test is used. The McNemar is a nonparametric test used to evaluate categorical data from matched pairs.

Kruskal-Wallis Test

Just as the ANOVA is a generalization of the t-test, the Kruskal-Wallis test is a generalization of the Wilcoxon rank sum test and serves as a nonparametric approach to the ANOVA. The Kruskal-Wallis test allows hypothesis testing when there are more than two samples and ordinal data.

Proportions

Proportions can be evaluated using the binomial distribution if there are only two possible outcomes, such as success and failure, or present and absent for simplicity. For example, a study might compare the percentage of individuals with breast cancer in a sample with the general population. When the number of trials is large, the binomial distribution is difficult to use. If the number of trials is moderately large and the probability of success in each trial is not too extreme in either direction, the central limit theorem applies, and a normal approximation to the binomial distribution can be used.

The Poisson test is used when uncommon conditions are considered because the expected number of events per unit of time follows a Poisson distribution. When using the binomial distribution, the focus is on a finite number of trials, in which the number of events is limited to the number of trials. Under the Poisson distribution, however, the potential number of trials is infinite and, as a result, the number of events can also be unlimited. As with the binomial distribution, when the expected number of events per unit of time is large, the Poisson distribution becomes difficult

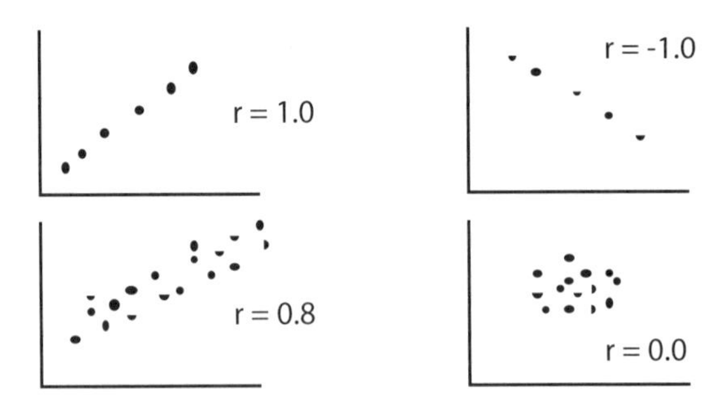

Figure 2. Variable correlation.

to use. When the number of expected events per unit of time is at least 10, the normal approximation to the Poisson distribution is used.

Correlation

Correlation measures the strength of association between variables (for example, the likelihood that A will increase or decrease while B is increasing or decreasing). Correlation methods show the general linear relationship between variables. Values of the correlation coefficient (r) vary from positive one to negative one (+1 to −1), where the respective extremes indicate perfect agreement and disagreement (Figure 2). Positive correlation values indicate that as one variable increases, the other does also. Conversely, a negative correlation indicates that as one variable increases, the other decreases. If $r = 0$, the variables are not linearly related, though the lack of an observed association may also reflect limits in the data collected. For example, if inclusion criteria are particularly restrictive, an artificial association—or the lack of one—may be observed. It is critically important to recognize that variables may be related without a causal relationship existing (i.e., correlation is not causation). Similarly, simply saying that variables are correlated is not informative. It is preferable to indicate how strongly the linear relationship is and in what direction.

There are several types of correlation measures, including Kendall rank correlation, Pearson product moment correlation, and Spearman rank order correlation. Kendall correlation measures the relationship between ordinal variables, Pearson correlation assesses the association between almost normally distributed continuous variables, and the Spearman method is used when at least one of the variables is not normally distributed.

Other types of correlation coefficients estimate the degree of agreement within or between raters. These measures are referred to as interclass or intraclass correlation. An example of this type of measure is the κ (kappa) statistic, which is used to describe the degree of agreement by several observers of the same subject. Specifically, if we are interested in how reproducibly a variable is measured by different surveys or different tools or observers, κ is used to compare the observed and expected probabilities of agreement between the different measurements. In general, κ is estimated using a one-tailed test, because negative values typically do not provide useful information.

When the correlation coefficient is squared (r^2), the result is the coefficient of determination. This value represents the percentage of the variance in the dependent variable that is explained by the independent variable. Note that in the context of a regression model, the r^2 increases with the number of covariates in the model, but building such a model is not a headlong pursuit of the highest r^2. To help construct a model that maximizes the variance in the dependent variable, but that is also relatively parsimonious, the adjusted r^2 is used. The adjusted r^2 penalizes the r^2 for possible over fitting, where there are too many independent variables for the given sample size, resulting in unreliable increases in r^2 that will not be confirmed with a larger sample size or future study. Researchers should be held accountable and follow a rule of parsimony.

Regression

Regression methods are used to estimate the relationship between variables. Simple regression describes the relationship between a single independent variable and the dependent variable; multiple regression is used when there is more than one independent variable. Additional variables of interest that may improve a model are also referred to as covariates. These methods provide an explanation and prediction of expected relations within the range of the data; for example, regression can be used to determine the ability of smoking to predict lung cancer. Extrapolation of model results outside the observed range of data is not recommended.

A variety of regression methods exist, including linear, logistic, survival, and Poisson models. Linear regression methods are used when a straight-line relationship is assumed between the dependent and independent variables. When the dependent variable is binomial such as cured or not cured, logistic regression is used. Survival data are analyzed using proportional hazards regression (and related) methods. Major assumptions underlie each of these models, though methods exist that provide some flexibility when assumptions are not met. For example, the logistic model typically requires the outcome to have only two possibilities, but the ordered logistic model can be used when the dependent variable has more than two categories. Similarly, the proportional hazards assumption is an important part of survival analysis, but when this assumption does not hold true, it may be useful to allow the effect of variables to vary over time (time-varying covariates).

Box 1. Linear Regression Model Assumptions

- **Existence:** For each value of the independent variable, the dependent variable is random, with a probability distribution that has finite mean and variance.
- **Independence:** Values of the dependent variable are statistically independent from each other.
- **Linearity:** The relationship between the mean value of the dependent variable, given the independent variables, and the independent variables is linear.
- **Homoscedasticity:** The variance of the dependent variable is the same for all levels of the independent variable.
- **Normality:** For any values of the independent variable, values of the dependent variable are normally distributed.

These assumptions are generally evaluated by examining histograms, scatter plots, and graphs of residuals, which are the distance between each observation point and the regression line, among other methods.

Linear Regression

Like all statistical methods, regression models are used appropriately only if certain assumptions are met. The assumptions underlying the linear regression model are shown in Box 1.

The general form of the linear regression line is as follows:

$$\mu\{Y/X\} = \beta_0 + \beta_1 X_1 + \ldots + \beta_n X_n$$

where $\mu\{Y/X\}$ refers to the mean of the dependent variable, given the values of the independent variable(s) $X_1 - X_n$; β_0 is the intercept of the line, measured in the same units as the dependent variable; and $\beta_1 - \beta_n$ are the slopes of the lines. Another name for β is coefficient. The slope of the line (or coefficient) is the change in the value of the dependent variable caused by a 1-unit change in the independent variable. For example, in a simple regression model, if the coefficient is 1.0, the value of the dependent variable increases by one for each 1-unit increase in the independent variable. However, this interpretation depends on the scale of the independent variable. If, for example, the independent variable has been transformed to its logarithm, the coefficient will still be the change in the mean of dependent variable for each 1-unit change in independent variable, but the unit change will be interpreted as a change on the particular log scale, such as a doubling or a 10-fold increase. In multiple linear regression, any given coefficient represents the change in the mean value of dependent variable for each 1-unit change in independent variable, assuming the other variables are held constant or adjusting for the other variables.

Dummy, or indicator, variables can be used when an independent variable has more than one level. For example, to examine the effect of sex on blood pressure, one way to code sex is as a single variable that has two different values, one for males and one for females.

The slope and the intercept (also called the constant) in a regression model can be evaluated by testing the null hypothesis that these terms are equal to zero. If we accept the H_0 that the slope is zero, this observation suggests that the independent variable does not contribute to explaining or predicting the dependent variable, or that the relationship between the parameters is not linear (in a linear model). When the H_0 is rejected, we generally conclude the opposite, although other models with linear components (curvilinear graphs, for example) may also fit the data well. Under the test for the intercept, if we fail to reject the H_0 that the intercept is zero, we can remove the intercept from the model. This step is analogous to forcing the line through the origin, but observations for some variables are generally unavailable when some independent variables, such as age, equal zero. As a result, we are usually not interested in the statistical significance of the intercept.

Logistic Regression

In logistic regression, the goal is to estimate the relationship between study variables and an outcome that can be categorized into two groups, such as disease or no disease, or severe pain or no severe pain. Generalized logistic methods, such as ordered logistic regression, allow modeling an outcome with more than two categories; however, this discussion focuses on the dichotomous-outcome logistic model of two groups of outcomes.

Like other regression models, a logistic regression model has β coefficients that indicate how much the dependent variable changes with a 1-unit change in the independent variable. In multiple logistic regression, the coefficient represents the relationship between each independent variable and the dependent variable when all other independent variables are held constant. The result of the logistic regression model is the OR, regardless of the type of underlying study. The OR quantifies the odds of the outcome in those exposed to the independent variable divided by the odds of the outcome (the dependent variable) in unexposed individuals. For example, an OR of 2.0 indicates that the independent variable is associated with a 100% higher risk of the outcome, whereas an OR of 0.75 indicates that the independent variable is associated with a 25% decrease in the likelihood of the outcome. In addition, when a CI for an OR includes 1.0, the effect of the independent variable is interpreted as not being statistically different from zero (i.e., there is no difference in the outcome between individuals who received the intervention and those who did not).

Of note, the interpretation of the β coefficients may differ on the basis of the type of variable used and how it is coded. When the independent variable of interest has only two possible outcomes—coded as 0 and 1, for example—the OR provides an estimate of the likelihood of the outcome between individuals with the variable equal to 1 and those with the variable equal to 0. When the independent variable has more than two possible categories, the OR is derived from comparing individuals with the variable equal to one level with individuals in the reference, or baseline, group. Similarly, when the independent variable is continuous, the OR represents the change in the likelihood of the outcome associated with a 1-unit increase in the value of the independent variable. Of note, the scale of the independent variable is very important. For example, a 1-mm mercury increase in blood pressure may not be clinically significant. However, if a variable is measured on a small scale, 1 unit may be far too large.

The number of statistical tests available, combined with the infinite number of ways a study can be designed, can be overwhelming for both the investigator analyzing the data and the clinician interpreting the results. Table 3 is organized to help the reader understand how many commonly used statistical tests fit in on the basis of the type of data used and the design of a study.

Survival Analysis

Survival analysis methods apply to questions in which the outcome is time until an event takes place. Events of potential interest may include progression of disease, recurrence of a condition, and death, among many others. One advantage of survival analysis techniques is that they account for incomplete observation, or censoring. Censoring refers to instances in which the investigator knows something about an individual's experience, but the exact time until the outcome is unknown. For example, an individual may withdraw from a study, be lost to follow-up, or not experience the event being measured by the time the study ends. In addition, survival data may be considered right or left censored. Right censoring occurs when the observation period ends without some individuals yet experiencing the event of interest. Left censoring occurs when the event of interest happens sometime before the start of observation. Observations may also be interval censored, which is when the event occurs at some unknown time between scheduled observation points. For example, if a subject's disease progresses sometime between the scheduled follow-up appointments at 3 months and 12 months, interval censoring has occurred. Another mechanism of incomplete observation is called truncation, but censoring is inherently different from truncation. Specifically, censoring occurs at the individual level, whereas truncation is a design issue. Thus, if observation does not begin until some specified time after the start of exposure, the data set is left truncated. If all people in a study experience the event before the study begins, the data set is right truncated.

Survival data are often analyzed using the Cox proportional hazards model and are reported as survival curves (Figure 3). The Cox

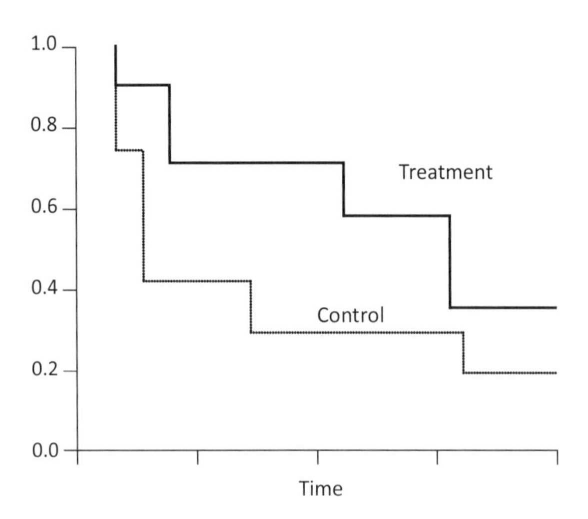

Figure 3. Survival curves.

proportional hazards model is a regression model that supplies an estimate of the hazard ratio, which in the clinical literature is commonly an estimate of the ratio of the hazard rate in the treatment versus control group. The hazard rate is the likelihood that an event will occur in the next time interval, divided by the length of the time interval. If the time intervals are very small, it is in essence an instantaneous rate; however, it is not an estimate of how quickly the event will occur. For example, the hazard ratio can report the odds of a patient healing quicker, but it does not indicate how much quicker. The main assumption underlying this model is the proportional hazards assumption. Although hazard generally refers to the chance that some event will occur, this assumption states that the hazard—defined as the instantaneous potential per unit of time for a person to experience the event of interest, given that the person survives until that time—is proportional to that for any other individual and that it is independent of time. The proportional hazards assumption can be tested statistically and by examining curves on a survival graph. For example, if survival curves cross, the proportional hazards assumption does not hold.

Methods for addressing cases when the proportional hazards assumption does not hold include examining the model more closely to determine which covariates contribute non-proportionally, stratifying on the exposure variable, or using an extended model in which some variables are modeled to permit them to vary over time.

Poisson Regression

Poisson regression models the number of events, with the following assumptions: that the incidence rate reflects how often events occur; that the incidence rate multiplied by the exposure provides an estimate of the expected number of events; that during very small exposure periods, the probability of more than one event taking place is small; and that nonoverlapping exposures are mutually independent. Coefficients from Poisson models represent the change in the log incidence rate for a 1-unit change in the independent variable.

SELECTED BIBLIOGRAPHY

Aaron SD, Fergusson DA. Exaggeration of treatment benefits using the "event-based" number needed to treat. CMAJ 2008;179:669–71.

Bland JM, Altman DG. Survival probabilities (the Kaplan-Meier method). BMJ 1998;317:1572.

Bulpitt CJ. Subgroup analysis. Lancet 1988;2:31–4.

Chatellier G, Zapletal E, Lemaitre D, Menard J, Degoulet P. The number needed to treat: a clinically useful nomogram in its proper context. BMJ 1996;312:426–9.

Cody RL, Slack MK. Crossover design in pharmacy research. Ann Pharmacother 1992;26:327–33.

Datta M. You cannot exclude the explanation you have not considered. Lancet 1993;342:345–7.

Egger M, Davey-Smith G. Misleading meta-analysis. Lessons from an "effective, safe, simple" intervention that wasn't. BMJ 1995;310:752–4.

Etminan M, Levine M. Interpreting meta-analyses of pharmacologic interventions: the pitfalls and how to identify them. Pharmacotherapy 1999;19:741.

Etminan M, Wright JM, Carleton BC. Evidence-based pharmacotherapy: review of basic concepts and applications in clinical practice. Ann Pharmacother 1998;32:1193–200.

Eysenck HJ. Meta-analysis and its problems. BMJ 1994;309:789–92.

Gibaldi M, Sullivan S. Intention-to-treat analysis in randomized trials: who gets counted? J Clin Pharmacol 1997;37:667–72.

Goodman SN. Stopping at nothing? Some dilemmas of data monitoring in clinical trials. Ann Intern Med 2007;146:882–7.

Hayden GF, Kramer MS, Horwirtz RI. The case-control study: a practical review for the clinician. JAMA 1982;247:326–31.

Hayward RA, Kent DM, Vijan S, Hofer TP. Reporting clinical trial results to inform providers, payers, and consumers. Health Aff 2005;24:1571–81.

Mueller PS, Montori VM, Bassler D, et al. Ethical issues in stopping randomized trials early because of apparent benefit. Ann Intern Med 2007;146:878–81.

Oxman AD, Sackett DL, Guyatt GH. Users' guide to the medical literature, parts I–VIII. JAMA 1993;270:2093.

Pocock SJ, Travison TG, Wruck LM. How to interpret figures in reports of clinical trials. BMJ 2008;336:1166–9.

Podrebarac T, Tugwell P, Hebert PC. A reader's guide to the evaluation of causation. Postgrad Med J 1996;72:131–6.

Porta MS, Hartzema AG. The contribution of epidemiology to the study of drugs. Drug Intell Clin Pharm 1987;21:741.

Rousseeuw PJ. Why the wrong papers get published. Chance 1991;4:41–3.

Sackett DL, Cook RJ. Understanding clinical trials: what measures of efficacy should journal articles provide busy clinicians? BMJ 1994;309:755–6.

Sacks HS, Berrier J, Reitman D, Ancona-Berk VA, Chalmers TC. Meta-analysis of randomized controlled trials. N Engl J Med 1987;316:450–5.

Sheiner LB, Rubin DB. Intention-to-treat analysis and the goals of clinical trials. Clin Pharmacol Ther 1995;57:6–15.

Sterne JAC, Smith GD. Sifting the evidence—what's wrong with significance tests? BMJ 2001;322:226–31.

West PM. Literature evaluation. In: Schumock GT, Brundage DM, Chapman MM, et al., eds. Pharmacotherapy Self-Assessment Program, 5th ed. The Science and Practice of Pharmacotherapy II Module. Kansas City, Mo.: ACCP, 2005:93–114.

Wiffen PJ, Moore RA. Demonstrating effectiveness—the concept of numbers-needed-to-treat. J Clin Pharm Ther 1996;21:23–7.

Yusuf S, Wittes J, Probstfield J, Tyroler HA. Analysis and interpretation of treatment effects in subgroups of patients in randomized clinical trials. JAMA 1991;266:93–8.

SELF-ASSESSMENT QUESTIONS

1. You have been asked to design a study in which the expected prevalence of a drug effect is very low. To compare the experience of individuals who took the drug with that of individuals who did not, which one of the following statistical tests is most appropriate to use?

   A. z-test
   B. Fisher exact test
   C. t-test
   D. Analysis of variance (ANOVA)

2. You want to compare the effect of long-term use of a drug on patient height. You compare results between groups of children ages 2–5, 6–10, 11–14, and 15–18 years. Which one of the following statistical tests is the most appropriate choice?

   A. Kruskal-Wallis
   B. Chi-square
   C. ANOVA
   D. Paired t-test

3. A study assessed the relationship between a set of clinical and demographic characteristics and pain in individuals at the end of life who received hospice care. Pain was assessed when the person began receiving hospice care and every few days until they died or were discharged. The outcome of interest was whether the patient had severe pain at his/her last observation. Which one of the following types of analysis would be best to use in this case?

   A. Linear regression
   B. Survival analysis
   C. Logistic regression
   D. Ordered logistic regression

4. In a study, the r^2 = 0.32. Which one of the following interpretations best explains this result?

   A. The model explains 57% of the variation in the outcome.

B. No conclusions can be drawn because it is not apparent whether the estimated coefficients for each covariate were statistically significant.
C. About 32% of the variation in the outcome was explained by the independent variable(s).
D. The model explains 75% of the variation in the outcome.

Questions 5 and 6 pertain to the following case.
A randomized, controlled trial assesses the effects of a treatment for multiple sclerosis on functioning in three groups of adults. Investigators assess global functioning with an ordered, Likert-type scale.

5. Which one of the following statistical tests is most appropriate to assess differences in functioning between the groups?
   A. Kruskal-Wallis
   B. Multiple ANOVA
   C. ANOVA
   D. Analysis of covariance (ANCOVA)

After the study is complete, the investigators decide to conduct subgroup analyses to identify groups most likely to benefit from the experimental treatment. Some results indicate that the treatment resulted in statistically significantly improved functioning for specific subgroups of study participants.

6. Which one of the following is the most appropriate way to report these post hoc results?
   A. Results from the post hoc analyses should not be reported.
   B. Results from the post hoc analyses should be reported in the same way as the analyses planned before starting the study.
   C. Results from the post hoc analyses should be adjusted for multiple comparisons before being reported.
   D. Results from the post hoc analyses should be reported only qualitatively.

7. In a study to assess the risk of stroke, myocardial infarction, or death in the first 30 days after carotid endarterectomy, individuals

who received the experimental intervention had an event rate of 8.2%, whereas the event rate among individuals in the control group was 3.7%. You have been asked by the Pharmacy and Therapeutics Committee to estimate how many people would need to be treated with this intervention for one person to experience the outcome. Which one of the following is the best interpretation of that information?

A. About 22 individuals would need to be treated with the experimental intervention for one person to be harmed by the intervention.
B. About 0.2 individuals would need to be treated with the experimental intervention for one person to benefit from the intervention.
C. About 12 individuals would need to be treated with the experimental intervention for one person to benefit from the intervention.
D. About 27 individuals would need to be treated with the experimental intervention for one person to be harmed by the intervention.

8. A study tests whether the use of a new drug adherence aid has an effect on congestive heart failure exacerbations. Researchers want to evaluate the proportion of patients with an exacerbation of their congestive heart failure symptoms in a group of people who used the adherence aid compared with a group of people who did not. To assess whether the changes observed are statistically significant, which one of the following tests is best?

   A. McNemar
   B. Chi-square
   C. Two-sample t-test
   D. Mann-Whitney U-test

9. A prospective, randomized, placebo-controlled trial of a new antidepressant drug reports that for the primary outcome of response rate (50% decrease in Hamilton Rating Scale for Depression), there is no difference between the drug and placebo ($p>0.05$). The researchers also report that they decided to do an additional previously unplanned

analysis of the data after the conclusion of the trial and that they were able to show a better response rate for the new drug versus placebo in the women in the trial (p=0.04). Which one of the following is the most valid conclusion from this trial?

A. The new drug works in women but not in men.
B. The trial should have listed two primary outcomes.
C. The response rate reported for the entire group of participants should be analyzed as a secondary outcome.
D. A prospective trial designed to test the drug in men compared with women should be considered.

10. The Women's Health Initiative study compared the use of conjugated equine estrogens plus medroxyprogesterone with placebo in healthy postmenopausal women and reported a higher number of cardiovascular events in women receiving this hormone replacement therapy regimen. Which one of the following correct descriptions of the results about cardiovascular disease reported in the trial would be best to use in discussions with patients?

    A. Cardiovascular events increased in women taking the active drug.
    B. Patients taking the drug had an increased risk of cardiovascular events that was statistically significant.
    C. For every 10,000 women who take the drug for 1 year, there will be seven extra cardiovascular events.
    D. The rate of cardiovascular events was increased by 29% in women taking hormone replacement therapy.

11. A trial compares drug X and drug Y for treating nausea and vomiting associated with pregnancy because clinicians believe there may be differences in their efficacy in preventing nausea and vomiting in pregnant women. Drug X has been used for many years and has a large evidence base showing efficacy and safety. Drug Y was recently introduced to treat nausea and vomiting associated with chemotherapy, but it has not been well studied in patients with nausea and vomiting caused by pregnancy. Patients will be randomized to one of these two drugs. The end point for this trial will be based on patients' ranking of their nausea and vomiting 3 hours after taking the drug to which they have been assigned. Nausea and vomiting will be graded

using the following scale: 0 = no nausea, 1 = mild nausea, 2 = moderate nausea, 3 = severe nausea, and 4 = vomiting. Which one of the following types of statistical tests is best suited to test for differences between these drugs?

A. Wilcoxon rank sum test
B. Wilcoxon signed rank test
C. Student t-test
D. Paired Student t-test

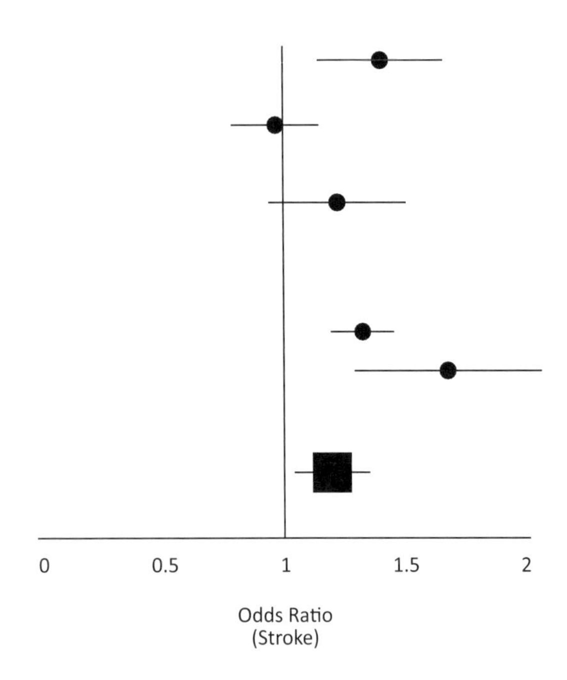

12. The results of a meta-analysis investigating the effects of a drug on stroke are shown in the figure above. Which one of the following is the best interpretation of these results, as portrayed in this figure?
    A. The drug being studied does not have an effect on the frequency of stroke.
    B. The drug being studied decreases the frequency of stroke.
    C. The drug being studied increases the frequency of stroke.
    D. The drug's effect on stroke cannot be determined from the figure.

13. A new antipsychotic drug was compared with a previously available drug in a prospective, randomized, and blinded trial. This 12-month trial measured the frequency of inpatient psychiatric admissions in both treatment groups. At the end of the trial, 6% of the patients taking the new drug had an inpatient psychiatric admission compared with 11% of the patients taking the older drug (p=0.03). In presenting the results of this trial to members of the formulary committee at the health maintenance organization where you work, which one of the following statements would provide the committee with the best information to use in deciding the formulary status of the new drug?

    A. The new drug decreased psychiatric admissions by 45%.
    B. The new drug decreased psychiatric admissions by 83%.
    C. You would need to treat two patients with the new drug to avoid one admission.
    D. You would need to treat 20 patients with the new drug to avoid one admission.

14. A trial studied an antihypertensive drug to assess its effects on blood pressure. Researchers compared the blood pressure of 150 patients at baseline and then again after the drug had been taken for 2 weeks. The results show that this antihypertensive drug lowers systolic blood pressure by an average of 9 mm Hg (p=0.04; two-sample t-test) and diastolic blood pressure by an average of 7 mm Hg (p=0.03; two-sample t-test). Which one of the following statements is consistent with these reported results?

    A. This drug is not effective at lowering blood pressure.
    B. This drug is effective at lowering only diastolic blood pressure.
    C. This drug is effective at lowering both diastolic and systolic blood pressure.
    D. This drug may or may not lower blood pressure; the results are unreliable.

Questions 15 and 16 pertain to the following case.

A linear relationship between the dosage of a new chemotherapeutic drug and pulmonary function is being investigated. Measurements of the forced expiratory volume in 1 second are collected as a measure of lung function and plotted against the corresponding dosage of drug each patient received.

15. Which one of the following is an appropriate statistical approach to assess any correlation between drug dose and forced expiratory volume in 1 second?

    A. Pearson product moment coefficient
    B. ANOVA
    C. Spearman rank correlation
    D. ANCOVA

16. The statistical test applied to the data studying the relationship between this drug and lung function reports an $r = -0.46$ ($p<0.05$). Which one of the following represents the best interpretation of these results?

    A. Seven percent of the variation in forced expiratory volume in 1 second is associated with the dose of this drug.
    B. Twenty-one percent of the variation in forced expiratory volume in 1 second is associated with the dose of this drug.
    C. Forty-six percent of the variation in forced expiratory volume in 1 second is associated with the dose of this drug.
    D. Ninety-two percent of the variation in forced expiratory volume in 1 second is associated with the dose of this drug.

17. A new laxative was compared with psyllium in adults and children between the ages of 2 and 65 years. The study assessed the length of time to the first bowel movement after taking one of the two study drugs. At the end of the trial, a statistical analysis of the outcomes resulted in a p value of 0.30. On the basis of this result, the researchers reported that they then decided to look for differences in effect between men and women and among different age groups (i.e., 2–5 years old, 6–12 years old, 13–18 years old, 19–55 years old, and older than 55 years). At the end of these analyses, the new laxative was found to provide superior relief of constipation in women older than 55 ($p<0.05$). Which one of the following is the best interpretation of these results?

A. This drug works better than psyllium only in women older than 55.
B. Until more trials are conducted, it should be concluded that this drug works no better than psyllium.
C. The researchers found no overall difference, so this drug cannot work better in certain subgroups.
D. Because this drug works better in women older than 55, it should work in all women.

18. A randomized, double-blind trial tests the hypothesis that a new vasodilator improves the symptoms of congestive heart failure. Patients are randomized to either the new drug or placebo and continue their existing congestive heart failure pharmacotherapy. At the conclusion of the trial, the intention-to-treat analysis shows no statistically significant difference between the groups. The authors then decide to perform an analysis based on the actual therapies the patients received. Data for patients who did not finish at least 60% of their assigned drug were evaluated as if these patients were in the placebo group. This analysis showed that the new drug was more effective than placebo ($p<0.05$) in lessening the symptoms of congestive heart failure that were measured. Which one of the following is the best course of action on the basis of these results?

    A. Recommend the new drug for all patients with congestive heart failure.
    B. Recommend the new drug for patients who adhere to their current therapies.
    C. Recommend the new drug only for early-stage congestive heart failure.
    D. Do not recommend the new drug; wait for further studies.

LITERATURE INTERPRETATION AND APPLICATION TO PATIENT CARE

Interpreting Results from Clinical Trials

INTRODUCTION

Interpreting scientific and medical literature is essential for improving health outcomes. To improve patient care, we must be able to determine whether what is reported in the literature is applicable to our patients (i.e., can the results be extrapolated to our patient population). The investigator should make every effort to avoid common methodological errors including relying on statistical software to make decisions, lacking attention to detail when collecting data, assuming that one statistical method fits all questions, and dumping poorly organized information into unclear charts and graphs. In addition, although post hoc analyses are useful for generating hypotheses, these reports should be distinguished from the results of a priori hypothesis testing. Even if the investigator has avoided common methodological errors, the reader must still interpret the literature and discern whether the results can be extrapolated to his or her patient or patient population. The following sections explain concepts important to interpreting the medical literature together with examples of applying these concepts to the cardiovascular literature.

OUTCOME VARIABLES

Continuous and Discrete Variables

The two main types of outcome variables are continuous (quantitative) and discrete (qualitative). Continuous variables are quantifiable on an infinite (or almost infinite) scale. Examples of continuous variables are weight, height, serum creatinine levels, and blood pressure. Such variables can take on any value within a given range. Variables that can take value from a finite or countable set are called discrete variables. Discrete variables, which are always integer values, are also known as categorical variables. Such variables can be dichotomous, as when a patient can be in one of two states (e.g., hospitalized or not, dead or alive, revascularized or not); ordinal, as when categories are ordered or ranked but the differences between categories are not quantifiable (e.g., the New York

Figure 4. Possible relationships between surrogate outcomes and clinical outcomes. In case A, the surrogate outcome is linked to the clinical outcome of interest. In case B, the surrogate outcome bears no relationship to the clinical outcome. *Adapted from Fleming TR, DeMets DL. Surrogate end points in clinical trials: are we being misled? Ann Intern Med 1996;125:605–13.*

Heart Association functional classification); or nominal, as when named categories do not have an implied ordering (e.g., blood type).

The categorical outcomes in cardiovascular clinical trials are usually dichotomous. Some continuous variables can be discrete, such as number of children. Sometimes, continuous variables are transformed into discrete form for analysis (e.g., transforming body temperature into an ordinal variable such as low, normal, or high temperature). The main advantage of continuous variables is that they are powerful (i.e., differences can be detected more readily).

Surrogate Outcomes

Surrogate outcomes are parameters thought to be associated with clinical outcomes (e.g., blood pressure reduction as a surrogate for stroke outcomes); as such, they can substitute for clinical outcomes (Figure 4). Often, these measures are part of the disease process (i.e., they are physiologic measures), and they may be attractive for use in clinical trials because they are often more responsive to change. Surrogate outcomes also may reduce the follow-up time requirement, reduce the sample size, and be easier to measure than clinical outcomes. Despite these potential advantages, surrogate outcomes are not without limitations, as described in the following.

Frequent premature ventricular beats used to be well known as markers of poor prognosis after myocardial infarction (MI). The Cardiac Arrhythmia Suppression Trial (CAST) showed that encainide and flecainide were effective in suppressing these premature ventricular beats, considered a surrogate outcome for mortality. Therefore, encainide and flecainide were expected to reduce mortality after MI. The CAST study was a multicenter, randomized, double-blind, placebo-controlled trial of flecainide and encainide versus placebo in patients with a history of MI and with frequent premature ventricular beats. A total of 1,455 patients were enrolled and observed for about 300 days. The trial was discontinued prematurely by the Data and Safety Monitoring Board because of increased mortality in patients treated with encainide or flecainide (4.5%) versus placebo (1.2%); this represents a relative risk (RR) of 3.6 (95% CI, 1.7–8.5). The CAST study is an important cardiovascular trial because it showed the pitfalls of surrogate outcomes.

The primary concerns with the use of surrogate outcomes are validity and the ability to detect adverse effects. Validity may be a problem, as shown in the CAST study, because the association of clinical outcomes with the surrogate is often incompletely understood and may be invalid. In Figure 4, arrow A shows the presumed relationship between the surrogate and the clinical outcome. However, the relationship between the surrogate and the clinical outcome may not be valid (arrow B).

The smaller sample sizes and shorter trial duration generally needed for surrogate outcome trials may reduce the ability of trials to detect rare but important adverse effects. As such, surrogate outcomes may be useful in the early development of a new therapy, but they should never be viewed as a replacement for clinical outcomes. Trials using surrogate outcomes can be considered complementary to developing an understanding of the efficacy of a new compound.

Composite Outcomes

Composite or combination outcomes are commonly used in clinical trials. Composite outcomes, which combine events (e.g., death, hospitalization), may be useful when outcome events occur less commonly. Alternatively, they may be used to reduce sample size requirements for a study.

The Heart Outcomes Prevention Evaluation (HOPE) study was a randomized trial to evaluate the efficacy of ramipril versus placebo in reducing cardiovascular events in patients at high risk of such events but without left ventricular dysfunction or heart failure. Data from several trials of angiotensin-converting enzyme (ACE) inhibitors in patients with heart failure suggest these drugs have a protective effect on cardiovascular events. The HOPE trial also included a vitamin E versus placebo treatment arm; however, only the ramipril treatment results are considered here. In this trial, 9,297 patients at high risk of cardiovascular events were randomized to either ramipril or placebo and observed for 5 years. The primary outcome measure was a composite of MI, stroke, or death from cardiovascular causes. Of the patients receiving ramipril, 651 (14%) reached this primary outcome compared with 826 patients (17.8%) in the placebo group (RR = 0.78; 95% CI, 0.70–0.86, $p<0.001$). The survival curves showed a separation of the groups after about 1 year of therapy. This trial changed our way of thinking about the use of ACE inhibitors to prevent coronary artery disease events.

Because of their ability to increase power and reduce sample size requirements, composite outcomes are increasingly used. For the HOPE trial, a 22% relative risk reduction (RRR) in the primary outcome of cardiovascular events (MI, stroke, or death from cardiovascular causes) is clinically important. To be interpretable, the events that make up a composite outcome should be pathophysiologically linked (in this case, along the spectrum of cardiovascular outcomes). It would have been difficult to interpret the 22% reduction in the HOPE study had the composite end point been MI and emergency department visits for asthma.

Composite outcomes should count only one event per patient; this is usually defined as the first event, but it may be the clinical event deemed most important. Consider a patient in the HOPE study who suffers an MI, is hospitalized, has a stroke, and then dies. All four events are adverse cardiovascular outcomes included in the composite; however, to count these many outcome events for a single patient (quadruple counting) would be misleading. A limitation of the use of composite outcomes is that it presumes all components of the composite to be equally clinically important. Another disadvantage is that composite outcomes provide little detail on the exact nature of the treatment benefits. However, an examination of the detailed outcome events from HOPE shows a clear reduction in each of the outcome components as well as the composite (Table 4). It is important to look for component outcomes reporting; however, when breaking down a composite outcome into its component outcomes, power is lost (i.e., sample size for each component decreases), and the likelihood of a type II error increases.

ABSOLUTE RISK, RR, AND NUMBER NEEDED TO TREAT

The terms *absolute* and *relative* are commonly used in the biomedical literature, typically when discussing the rate of some event. For example, in a clinical trial comparing the effects of a new drug with placebo, the RRR is estimated by subtracting the percentage of individuals in the treatment group who have the outcome from the percentage of individuals with that event from the control group, divided by the percentage of individuals who have the event in the control group. In contrast, the absolute risk reduction (ARR) is simply the numerator of the above ratio (i.e., the percentage of individuals in the control group with the outcome, less that in the active comparator group). Relative measures are often larger than absolute measures and, as a result, are more commonly reported in the literature. Yet a large relative reduction may translate into few events, and absolute measures may be more meaningful to consumers and purchasers of health care. Similarly, relative measures are generally viewed as more relevant for etiologic questions, whereas absolute measures are more applicable for policy questions. This observation makes some sense because it is easy to imagine that employers and payers will likely be interested in the absolute number of events such as injuries, cases of disease, or missed days from work.

Table 4. Results from the Heart Outcomes Prevention Evaluation (HOPE) Study

Outcome	% RRR (95% CI), p value	NNT
Myocardial infarction, stroke, or death from cardiovascular causes (primary composite outcome)	22 (14–30), <0.001	26
Myocardial infarction	20 (10–30), <0.001	42
Stroke	32 (16–44), <0.001	67
Death from cardiovascular causes	26 (13–36), <0.001	50

CI = confidence interval; NNT = number needed to treat; RRR = relative risk reduction.

Information from Yusuf S, Sleight P, Pogue J, Bosch J, Davies R, Digynias G. Effects of an angiotensin-converting-enzyme inhibitor, ramipril, on cardiovascular events in high-risk patients. The Heart Outcomes Prevention Evaluation Study Investigators. N Engl J Med 2000;342:145–53.

When the ARR is expressed as a decimal (i.e., 1% = 0.01), its inverse is called the number needed to treat (NNT). This estimate refers to the number of individuals who must receive a treatment for some amount of time to prevent one undesirable outcome or achieve one good result. For example, in the Oxford League Table of Analgesic Efficacy in acute pain, the NNT to provide 50% or greater acute pain relief is listed for a variety of drugs. An analogous measure, the number needed to harm, refers to the number of individuals who must receive a treatment to cause one death or other serious injury. The number needed to harm is calculated in exactly the same way as the NNT, except that the outcome being considered is undesirable. Although a small NNT indicates that a drug is highly efficacious, a large number needed to harm is preferable because it indicates greater safety.

EXAMPLE OF QUANTIFYING TREATMENT EFFECTS IN CLINICAL TRIALS

Modern pharmacy practice is (or should be) driven by evidence. As such, some basic skills are required to interpret the increasingly complex medical literature. In this section, the Scandinavian Simvastatin Survival Study (4S) is used as an example to explore the ways in which treatment effects are quantified in clinical studies.

When the 4S trial was conceived, the evidence linking cholesterol lowering through the use of statins with mortality reduction was not well established. This study was designed to test the hypothesis that cholesterol lowering with simvastatin would improve survival in patients with coronary disease. A total of 4,444 patients with coronary disease and serum cholesterol concentrations of 5.5–8.0 mmol/L were randomized to receive simvastatin or matching placebo in a double-blinded fashion. Patients were observed for a median of 5.4 years. In the placebo group, 256 (12%) of 2,223 patients died, compared with 182 (8%) of 2,221 in the simvastatin group.

How well did simvastatin work? Several methods exist to quantify the effect size of a treatment. The simplest is to look at the difference in event rates between the placebo and active treatment groups. In this case, it is the difference between 12% mortality (placebo group) and 8% mortality (simvastatin group). This difference (in this case, 12% − 8% = 4%) is the ARR. The ARR is often difficult to interpret because it depends on the

baseline risk of the control or placebo group. Intuitivel
means something different when the baseline rate is high (e.g.
of 50% in the control group and 46% in the intervention group) ve
(e.g., event rates of 12% in the control group and 8% in the interve
group), as in the 4S trial.

Hence, treatment effects are commonly expressed as RRs. The RR is the event rate in the treatment group in relation to (i.e., divided by) the event rate in the control group. Thus, in the 4S trial, the RR is 8%/12% = 0.67 (≈0.70), meaning the event rate in the intervention group is 0.70 times (or 70%) the event rate in the control group. Although more easily interpreted than the ARR, the RR is not easily grasped. For example, retailers do not advertise a big sale as "all items 0.70." Instead, the sale is usually expressed as "30% off." This expression is analogous to the RRR, which is calculated by subtracting 0.70 from 1.0 (RRR = 1 − RR). In the 4S trial example, the RRR is 30% (1 − 0.70 = 0.30), stated as "simvastatin reduced mortality by 30%."

Even the RRR may have different interpretations depending on the baseline rate of events. If the baseline or control event rate is expected to be high (e.g., coronary events), a 30% reduction will have a very large clinical impact. However, if the baseline event rate is expected to be low (e.g., an 0.0008% incidence of developing amyotrophic lateral sclerosis), a 30% reduction is not as exciting. Continuing the example of a big retail sale, a 40% reduction (or RRR) in an item costing $1,000 would produce a savings of $400, which is more exciting than a 40% reduction on a $1 item (a savings of 40 cents). Therefore, an expression of impact that considers the baseline risk of the patient is required. The NNT is just such an expression of impact: the number of patients needed to treat to prevent one event. The NNT is calculated as 1/ARR. Using the 4S trial example, the NNT is 25 (1/0.04 = 25), meaning that treatment of 25 patients for about 5 years (the follow-up period in the 4S trial) will prevent one death. A reasonable NNT depends mainly on the risk and benefit of treatment. By convention, NNTs lower than 50 indicate treatments that are considered worthwhile. Therefore, the NNT of 25 in the 4S trial is reasonable.

Was the observed treatment effect "real," or was it a chance finding? Because a clinical trial can never enroll all patients with a given condition, it enrolls a sample from the pool of patients of interest, meaning we can never be 100% certain that the results reflect the entirety of patients with

ere is a risk of error in estimating the effect size
ntion, an error of 5% (0.05) or less is generally
tional (but arbitrary) p value less than 0.05. The p
y of obtaining the observed result by chance alone.
n, results with a p value less than 0.05 are considered
gnificant. In the 4S trial, the RR was reported as 0.70 with a p
.0003 (statistically significant). Note that statistically significant does not always mean clinically significant, which is an entirely different concept.

The RR and RRR are point estimates of effect size. In the 4S trial, a 30% RRR represents the best estimate of a treatment effect. Such an estimate is associated with a certain amount of error because it is calculated from one sample taken from the universe of patients with that condition; a different sample could lead to a different estimate. As such, it is important to know the precision of the estimate of the treatment effect. The 95% CI represents the range of the true treatment effect. In the 4S trial, the RR of death in the simvastatin group was 0.70 with a 95% CI of 0.58% to 0.85%. This means that with 95% certainty, the true treatment effect of simvastatin is between 0.58% and 0.85%. Expressing it as the RRR and subtracting these values from 1 gives an RRR of 30% with a 95% CI of 42% to 15%. By convention, the smaller number is stated first, so this is expressed as a 95% CI of 15% to 42%.

In a positive study such as the 4S, the true effect of simvastatin could be as large as a 42% reduction in mortality or as little as a 15% reduction in mortality. The lower level of the 95% CI (i.e., 15%) is an indication of the robustness of the finding (assuming that a 15% reduction in mortality is clinically important). The 95% CI provides more useful information than the point estimate of the RRR (30%) or the p value.

Had the 4S trial produced an RRR with a 95% CI that was −15% to 42% (i.e., crossing or including zero), the p value would have been greater than 0.05, and the finding would not have been statistically significant. Had the 4S trial resulted in the same RRR of 30% but with a 95% CI of −15% to 42%, it would have been declared a negative study, meaning that the treatment did not significantly reduce mortality. In a negative study, the 95% CI is still very useful. In this instance, the upper limit reflects that an RRR of up to 42% in mortality is possible. However, the lower limit of −15% indicates a possible increase in the RR of mortality for the treatment group.

Although point estimates of treatment effect are eas
95% CI is much more informative when making decisions a
ficacy of a treatment. Hence, the 95% CI should be evaluated wh
treatment's efficacy is assessed.

POWER AND SAMPLE SIZE

In evaluating a new treatment, the goal is to make statistical inferences about a certain group of patients (e.g., all patients with coronary artery disease). It is not feasible to study everyone with coronary artery disease, so a sample of the population (patient group) is drawn to make inferences that can be generalized. A key factor in making correct conclusions about treatment effect is drawing enough patients in the sample, which is where the concept of power enters.

The H_0 for most studies is that there is no difference between groups. Power is defined as the probability of correctly rejecting the H_0 (and concluding that there is a difference) when there really is a difference between treatment and control. Power is the ability of the statistical test to detect a difference when a difference exists. This is analogous to the requirement for adequate power on a microscope to differentiate between cellular structures. By convention, 80% power or greater is considered acceptable. Power is calculated as $1 - \beta$, where β is the probability of making a type II error (concluding there is no difference between treatment and control when there actually is). Conversely, α is the probability of making a type I error (i.e., saying that there is a difference between treatment and control when none actually exists). Conventionally, α is set at 0.05, which corresponds to the ubiquitous p less than 0.05 as indicating statistical significance.

Putting this information in practice, say the new drug "niastatinzetimibe" purports to reduce low-density lipoprotein (LDL) cholesterol. Early trials suggest it can reduce LDL cholesterol by 25% (an effect size of 25%). Previous clinical trials show the average LDL cholesterol to be 4.87 ± 1.0 mmol/L (SD). To design a randomized, controlled trial with 80% power and a standard α level of 0.05, the sample size for the study would have to be 22 (11 in each group). Figure 5 illustrates how the effect of treatment size influences sample size.

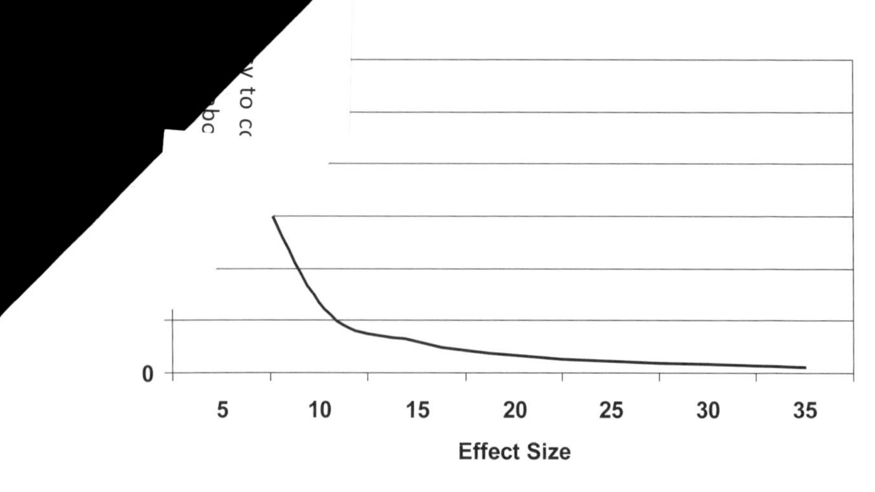

Figure 5. Relationship between effect size and sample size at a constant power and α.

On the basis of statistical calculations, if the new drug were expected to reduce LDL cholesterol by 35%, the sample size would drop to only eight (four in each group). However, if the drug were likely to produce only a 10% reduction in LDL cholesterol, the sample size would increase to 134 (67 in each group). In general, large effect sizes are easier to see—therefore requiring a smaller sample—whereas smaller effect sizes require a larger sample.

Now, what would be the effect on power if available funding limited enrollment to only 14 patients, assuming an effect size of 25% reduction, as above? Figure 6 shows that power decreases with smaller sample size. Therefore, with 14 patients, power drops to 62%, meaning there is now only a 62% chance of seeing a true difference. Conversely, increasing sample size is a way of increasing the power of a study.

Proceeding with a phase III mortality study for niastatinzetimibe, the 4S trial showed that the mortality rate in the control group would be about 12%. To test the assumption that niastatinzetimibe would reduce mortality by around 30% (i.e., down to 8.4%), a sample size to achieve 80% power at an α level of 0.05 would require 1,746 patients (873 per group).

What if the mortality rate in the patient population were lower than 12%? Figure 7 shows the effect of this mortality rate on the required sample size while maintaining an effect size of 30%. If the event rate were only 8%, the sample size would increase to 2,720 patients. If the event rate

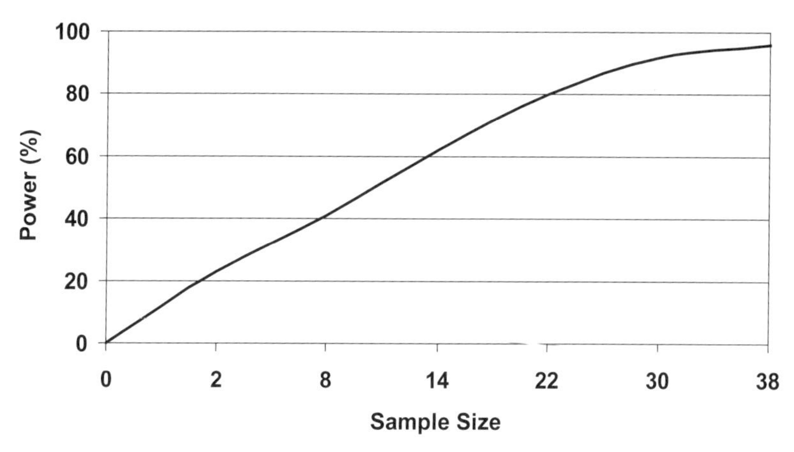

Figure 6. Relationship between sample size and power at a constant effect size and α.

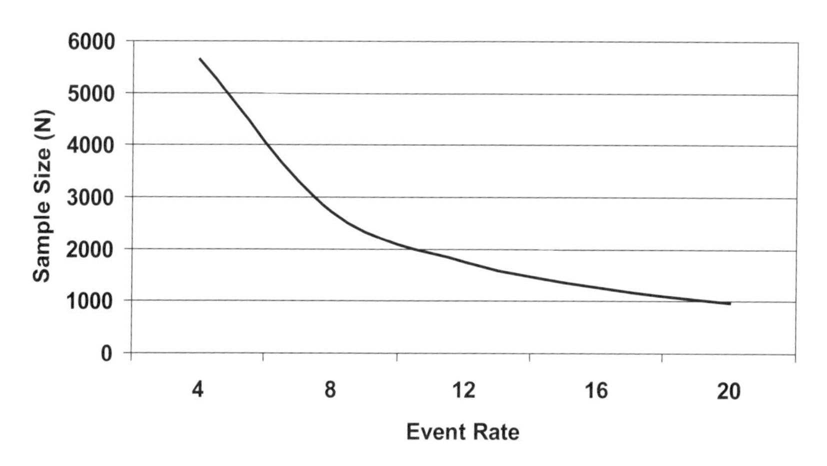

Figure 7. Relationship between event rate and sample size at a constant effect size, power, and α.

were only 4%, the sample size would swell to 5,640 patients, illustrating that discerning the treatment effects on rare events is much more difficult than when events occur more commonly.

If this study had been conducted with the previous sample size of 1,746 patients and the event rate had been less than 12% in the control group because of chance or poor planning, the effects on power would have been deleterious. An event rate of only 8% would drop the power of the statistical test to 64%, and an event rate of only 4% would drop the power to 40%.

A final area in which trial design can err is overestimating the effect size. Using the previous example, with an effect size of only 20% (rather than 30%), and keeping the sample size at 1,746 patients, the power drops to 49%. Although it is understandable that enthusiastic investigators would be overly optimistic about an effect size, an underpowered study would be created. When a neutral trial is reviewed, it is important to consider power. The result could have occurred because the trial was underpowered to show a difference, or there may actually have been no difference, even had the study been adequately powered.

ANALYSIS SELECTION

Statistical analyses allow us to determine how various factors affect outcomes. The t-test is the simplest form of unadjusted statistical analysis for the difference between two groups in a continuous outcome (e.g., blood pressure, cholesterol). The chi-square test is used for the difference between two groups in a dichotomous outcome (e.g., hospitalization, acute coronary syndrome, mortality).

But what about when there are many factors (independent variables) that might affect the outcome (the dependent variable)? When more sophisticated forms of statistical analyses are required to test such associations, the appropriate choice depends on the type of dependent (outcome) variable (continuous or dichotomous) and the type of independent variables (continuous or categorical). Figure 8 illustrates the tests that can be used for different types of dependent and independent variables.

To evaluate the effects of dichotomous independent variables on a continuous dependent (outcome) variable, an ANOVA is the commonly used statistical technique. Linear regression can also be used in this case. For continuous independent variables or a combination of both continuous and categorical independent variables on a continuous dependent variable, linear regression is used. The evaluation of a single variable versus several independent variables dictates the use of simple or multiple linear regressions, respectively. When the dependent (outcome) variable is dichotomous, the appropriate regression analysis is logistic regression.

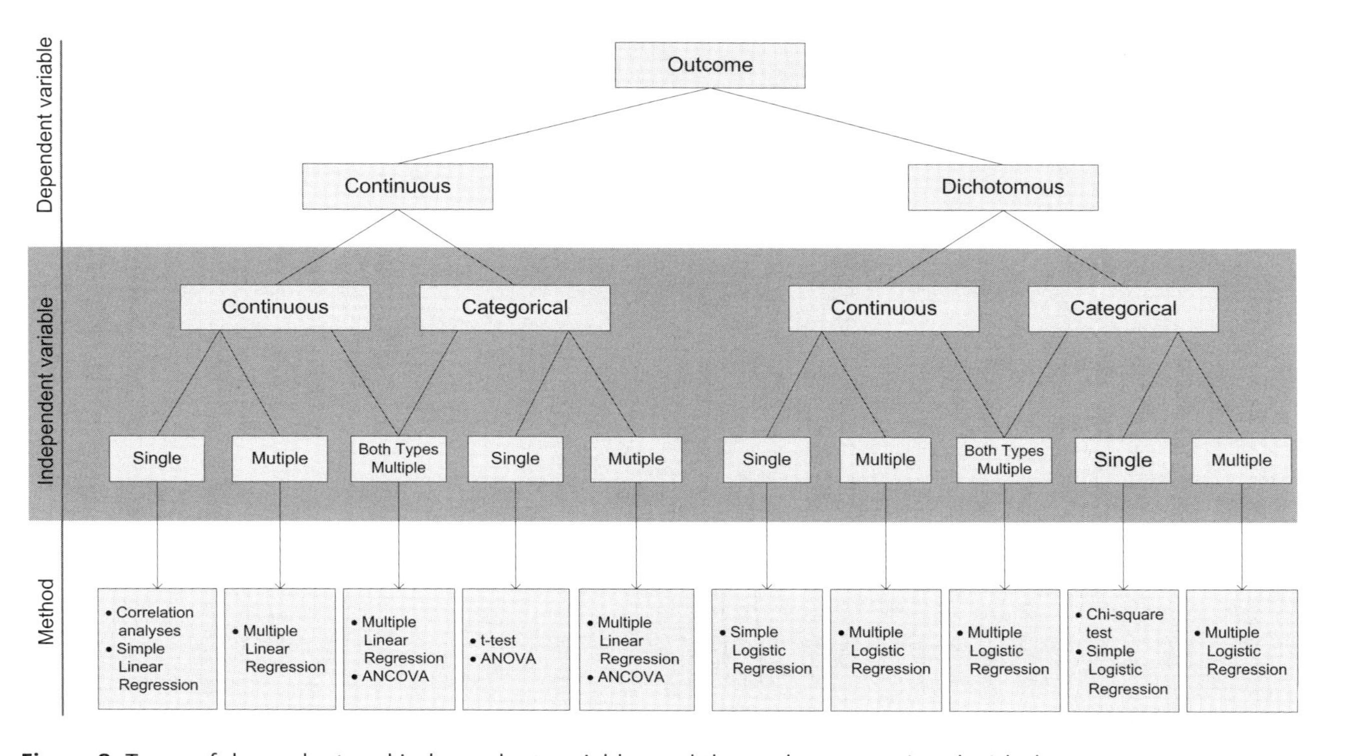

Figure 8. Types of dependent and independent variables and the analyses associated with them. ANCOVA = analysis of covariance; ANOVA = analysis of variance.

Regardless of the type of regression analysis used, the concept and objective are the same (i.e., to quantify and predict how the independent variables will affect the dependent variable). The formula is as follows:

$$Y = a + bX + E \text{ (e.g., mortality} = a + b \times \text{diabetes} + E)$$

where Y is the dependent (outcome) variable, a is the y-intercept (i.e., the value of Y when $X = 0$), b is the slope of the regression line (also called the regression coefficient), X is the independent variable, and E is an error term (or residual value). The error term is simply the effect of variables that are not included in the model.

There are several variations on these equations, depending on the relationship (e.g., straight line, curve, logarithmic) and the number of independent variables being tested; however, all variations predict what will happen to Y (outcome) with the change in X.

The Framingham Heart Study is a classic example of how regression analysis can be applied to an observational study. The study began in 1948 as an observational study of cardiovascular disease in a general population in Framingham, Massachusetts. In this case, the dependent variable was coronary heart disease. Several independent variables were evaluated: age, cholesterol, systolic blood pressure, weight, hemoglobin, cigarettes smoked, and electrocardiographic abnormalities. For the incidence of coronary heart disease, the study found a 30-fold difference in men and a 70-fold difference in women between the highest- and lowest-risk groups. The Framingham Study also showed that the strongest modifiable predictors (risk factors) for heart disease were cholesterol, cigarette smoking, and systolic blood pressure.

Regression analyses, unlike the simpler t-test or chi-square test, are useful in clinical trials when the analysis calls for an adjustment for the effects of other factors (confounders) that might also affect the outcome. In a randomized, controlled study, for example, the main intention is to evaluate the effect of a treatment. However, baseline discrepancies in patient characteristics (e.g., age, sex, comorbidities) may cause patients in the two groups to react differently to the treatment. The regression analysis can be adjusted for the effect of such variables by simply adding them as covariates in the regression model. A formula to estimate the effect of diabetes on mortality, adjusted for age, appears as follows:

$$Y = a + bX + cX2 + E \text{ (e.g., mortality} = a + b \times \text{diabetes} + c \times \text{age} + E)$$

SURVIVAL ANALYSES

Often, when evaluating the efficacy of a given treatment or intervention, preventing events (e.g., death, MI, revascularization, rehospitalization, other dichotomous outcomes) is of primary interest. Reporting the percentage of events in the intervention versus control group is fine; however, it does not give potentially important information on the timing of events. Survival analysis can provide important insights on the time course of a treatment benefit by studying the time between a patient's entry in a study and a subsequent event. Survival analysis is also useful when subjects have been exposed to the experimental treatment for varying lengths of time, as occurs in most clinical trials in which patients are enrolled during a certain period (the recruitment phase) and observed to a common closeout date.

The Metoprolol CR/XL Randomised Intervention Trial in Congestive Heart Failure (MERIT-HF) study is one of the classic pivotal trials of β-blockers in the treatment of heart failure. In this study, 3,991 patients with symptomatic heart failure and an ejection fraction less than 40% who were receiving standard drug therapy were randomized to receive metoprolol CR/XL or matching placebo. The primary outcome was all-cause mortality. The independent monitoring committee terminated the MERIT-HF study early because of the benefit of metoprolol. At an average follow-up of about 1 year, all-cause mortality was 7.2% in the metoprolol group compared with 11.0% in the placebo group. This translates to an RR of 0.66 (95% CI, 0.53–0.81) or an RRR of 34% (95% CI, 19%–47%). The NNT to prevent one death was 26 patients (1/[0.11 - 0.072]).

A 34% reduction in mortality and an NNT of 26 seem impressive, but there is more to the story, including the element of time, which is where the survival curve comes in. Survival curves show the percentage of people surviving at a particular time during the study, with the y-axis as percent survival and the x-axis as time. The cumulative incidence curve for mortality for MERIT-HF is shown in Figure 9; because the y-axis is cumulative mortality (the opposite of percent survival), this graph is a mortality (hazards) curve, simply the inverse way of presenting survival data.

The lines of the graph in Figure 9 look jagged because each event (death) is added onto the mortality (i.e., each zig in the line represents an event). The x-axis is time. Inspection of these curves can produce a better understanding of metoprolol's effect on mortality. It is evident that

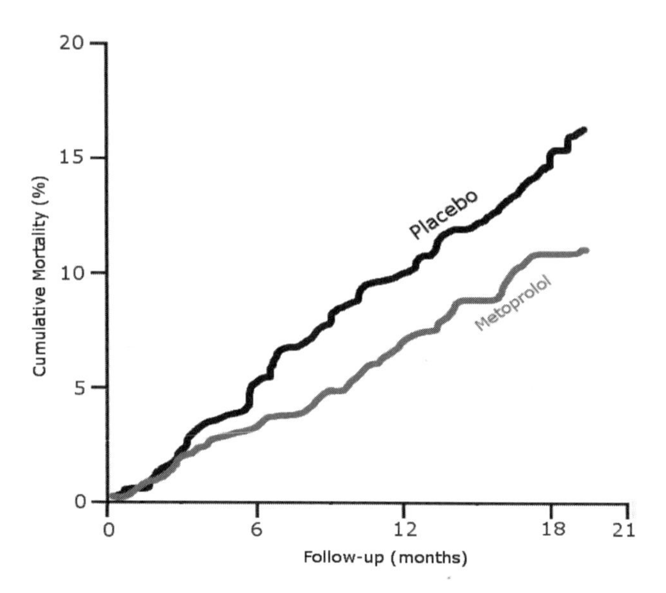

Figure 9. The cumulative incidence curve for mortality.

Reproduced with permission from: No authors listed. Effect of metoprolol CR/XL in chronic heart failure: Metoprolol CR/XL Randomised Intervention Trial in Congestive Heart Failure (MERIT-HF). Lancet 1999;353:2001–7.

after about 3 months, the line representing mortality in the metoprolol group begins to diverge and is lower than for placebo. It is also apparent that the distance between the curves appears to widen as time passes. Qualitatively, it appears that the benefits of metoprolol in patients with heart failure begin to accrue after 3 months of treatment and continue thereafter. A comparison of these two curves provides much more information about the time course of the outcome.

Compare this mortality curve with the graph of the 4S trial (Figure 10), which shows that the simvastatin and placebo curves do not begin to separate until about 18 months. This valuable insight into the time course of the benefit of simvastatin on mortality probably represents an effect on atherosclerotic lesions, which may take longer to manifest than the neurohormonal effects of metoprolol.

Another advantage of survival analysis is its help in quantifying the prevention or delay of an event (i.e., event-free survival). The disadvantages of using survival analysis include the amount of follow-up time

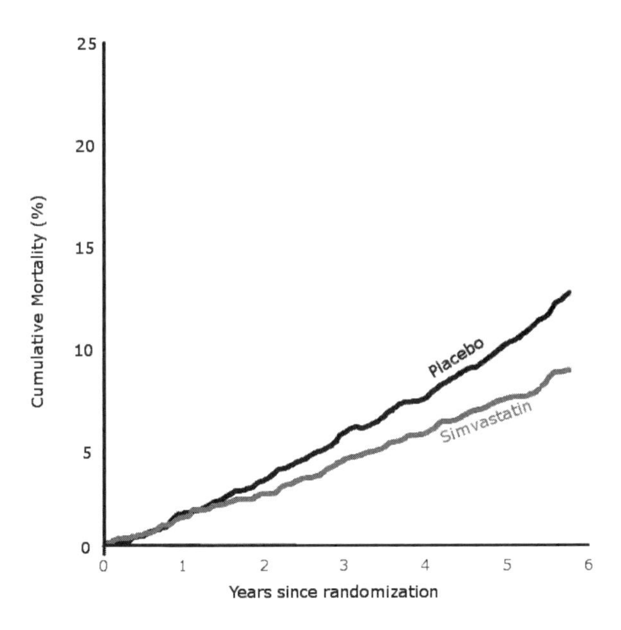

Figure 10. Mortality curve from the Scandinavian Simvastatin Survival Study.

Reproduced with permission from: No authors listed. Randomised trial of cholesterol lowering in 4444 patients with coronary heart disease: the Scandinavian Simvastatin Survival Study (4S). Lancet 1994;344:1383–9.

required to produce reasonable survival curves. In addition, the right-hand side of survival curves can be misleading because at the extremes of follow-up, there are often only a few patients contributing those events, possibly causing the curves to be skewed (i.e., the denominator is markedly lower at longer follow-up times). Of importance, look at the number of patients who have undergone follow-up at the far right-hand side of the graph (sometimes shown in the numbers under the x-axis indicating the number of patients at risk).

In summary, survival analysis is useful for dichotomous outcomes because it provides information on the dimension of time in relation to exposure to the treatment (and control). This can be particularly useful when patients have varying durations of follow-up time (to make use of all the information they can provide) as well as when the investigator is providing information on the delayed-onset effects of a treatment (as for cholesterol lowering and mortality in the 4S trial).

NON-INFERIORITY TRIALS

Clinical trials usually are intended to show that a new treatment is better than existing treatments or placebo; this is the basis for the traditional *superiority trial*. But what if the new treatment is very similar to an existing efficacious treatment and subjecting patients to a placebo group would be unethical? An example is testing the efficacy of a new antiretroviral drug for the treatment of HIV-1 (human immunodeficiency virus type 1) infection. In this case, the relevant clinical question is whether the new treatment is as good as the established treatment. Because it is impossible to prove that two things are exactly equivalent, the best alternative is to test that two treatments do not differ by more than some amount, usually referred to as delta (Δ). These studies are called *non-inferiority trials*. The new treatment A can be considered clinically acceptable if it is no worse than the existing treatment B by a factor of Δ. In a superiority trial, the H_0 is that treatment A is no different from treatment B; in conducting the study, the investigator hopes to reject this H_0 (Table 5). In contrast, in a non-inferiority trial, the H_0 is that A is better than or equal to B plus Δ. In this case, the alternate hypothesis is that A is no worse than B by a factor of no more than Δ.

The Ongoing Telmisartan Alone and in Combination with Ramipril Global End point Trial (ONTARGET) evaluated telmisartan, ramipril, or their combination in patients at high risk of vascular events using a non-inferiority trial design. Patients were randomized to receive ramipril, telmisartan, or both drugs in combination and were observed for a median of 56 months. The primary composite outcome was death from cardiovascular causes, MI, stroke, or hospitalization for heart failure. The primary outcome occurred in 16.5% of the ramipril group compared with 16.7% of the telmisartan group (RR = 1.01; 95% CI, 0.94–1.09). The combination therapy (ramipril and telmisartan) group showed similar outcomes but a higher rate of adverse events. Based on a non-inferiority margin

Table 5. Superiority and Non-inferiority Hypothesis Testing

Trial	Null Hypothesis	Alternate Hypothesis
Superiority	H_0: A = B	H_A: A ≠ B
Non-inferiority	H_0: A ≥ B + Δ	H_A: A = B + Δ

(Δ) of 13%, the authors concluded that telmisartan was non-inferior to ramipril (i.e., telmisartan was not more than 13% worse than ramipril on the outcome).

The ONTARGET study was set up to determine whether telmisartan would be non-inferior to ramipril on the basis of the incidence of cardiovascular-related deaths of patients receiving telmisartan being above that of patients receiving ramipril by 13% or less. This trial was conducted in the HOPE study setting, in which ACE inhibitors had been proved to reduce cardiovascular events compared with placebo. The question remained whether angiotensin receptor blockers could do the same. However, because it would be unethical to withhold ACE inhibitors from patients at high cardiovascular risk, a trial of telmisartan versus placebo would not be possible. Therefore, the non-inferiority design was appropriate.

How Δ is chosen is often on the basis of a minimally clinically important effect. For example, a difference of 7% in mortality might be stated as not that important; or, it might be stated that A would be acceptable if it would lead to no more than a 7% increase in mortality compared with B.

Although non-inferiority study designs can be useful when placebo controls cannot be used for ethical reasons, several characteristics make non-inferiority designs weaker than superiority designs. To avoid potential pitfalls, the non-inferiority design must include:

- A control treatment that has been proved efficacious. If this is not the case, an investigator may be placed in the awkward position of proving that two drugs are equally useless.
- The study population and outcomes must be similar to those of previous trials that proved the efficacy of the active control.
- Both treatments must be applied in an optimal fashion. (Potential problems include non-equipotent doses, low adherence, incomplete follow-up, cointerventions, and lack of blinding.)
- The trial must be appropriately powered. (Most non-inferiority designs require a much larger sample size than a corresponding superiority trial.)

Many superiority trials with indeterminant (negative) results are misinterpreted as non-inferiority trials. Worse yet, some authors try to salvage an indeterminant superiority trial by attempting to transform it into a non-inferiority trial.

HANDLING MISSING DATA

For many reasons, data are often missing or unusable. Study participants may not understand a question, or they may refuse to answer. Data collection may rely on different people and, as a result, vary in completeness. Unexpected events may prevent a person from completing a questionnaire or survey. If missing information obscures data and relationships important to the analysis, bias is introduced, which can result in the use of significantly hampered data.

To better understand the effect of missing data, the patterns and mechanism of missingness must be understood. Patterns of missing data refer to which variables and values are present and which are not. For example, in univariate missingness, data for a single variable are unavailable. A related pattern of missingness, called unit and item nonresponse, occurs when study participants do not complete a survey. Longitudinal studies can suffer from attrition or other loss to follow-up. Drug safety studies can provide an example of this pattern of missingness. When a large amount of data is missing, some variables may not be observed together. This limitation results in an inability to estimate the association between affected variables, or, even if estimates can be derived, they may be misleading. Finally, when variables are not observed at all, missing-data problems may be present.

In addition to the patterns described above, mechanisms that result in missing data are important to consider. The central issue is whether the values in the database are related to missing data. If missingness does not depend on the values of the data, missing or observed, the mechanism is called missing completely at random. Despite its name, this mechanism does not imply that the pattern of missingness is itself random, just that it is independent of the data. The next step on this continuum is called missing at random, and this mechanism applies when the missing data depend on the data values that are observed, not on those that are unobserved. The third mechanism is called not missing at random, and it applies when the missing data depend on the missing values. When data are missing completely at random, the observed data represent a random sample of all the data. However, when the missing data are not missing at random, the subsample analyses for the missing variables may be biased. An example is when final clinical measures are missing because of death. Simply ignoring the missingness will bias the findings toward more favorable outcomes. The ANOVA methods are not well suited to dealing with missing data; hence, when data are missing, regression models are preferred. However, good study design and conduct are the first line of defense against missing data. Analytic strategies are a distant second.

SUBGROUP ANALYSIS

Although a useful tool for obtaining more information from a data set, subgroup analyses should be viewed with skepticism when making qualitative inferences. There are several risks when performing subgroup analysis including accounting for the effect of multiple comparisons on the p value, loss of power, misclassification, and inability to assess for interactions. Finally, interpreting results from post hoc subgroup analysis that are data driven (i.e., comparisons suggested after reviewing the data) should only be viewed as hypothesis generating.

INTENTION-TO-TREAT ANALYSIS

Intention-to-treat analysis compares outcomes on the basis of initial group assignment, which allows determination of effect under conditions that are more reflective of real-life clinical practice. Compared with intention-to-treat analysis, per protocol analysis provides more useful information about method effectiveness (i.e., how well an intervention will work) and allows more generous estimates of treatment group differences; however, per protocol analysis is problematic if adherence is related to prognosis. As-treated analysis is similar to per protocol analysis; however, as-treated analysis does not eliminate data on the basis of adherence, which should result in less loss of power. As-treated analysis, however, jeopardizes the integrity of randomization: patients may end up switching groups because of some underlying, systematic bias.

SYSTEMATIC REVIEWS AND META-ANALYSES

A systematic review is a tool to summarize the medical literature. A systematic review that uses quantitative methods to summarize literature is called a meta-analysis. In contrast to the traditional narrative review, systematic reviews have an a priori defined protocol that includes the methods of data searching, evaluation, extraction, and combining. Narrative reviews are unsystematic (no defined process), and they often mix together opinions and evidence, which can lead to bias. Systematic reviews, when conducted using the results of properly designed, randomized, controlled trials, represent the highest level of causal inference (i.e., the highest level of evidence) and are extremely helpful in clinical decision-making.

A 2008 systematic review evaluated the impact of pharmacist care on patients with heart failure. The authors began by conducting a broad search of nine different electronic databases and included all randomized, controlled trials that evaluated the impact of pharmacist care on patients with heart failure. Outcome measures were defined as all-cause hospitalization, hospitalization for heart failure, and mortality. Of 3,115 studies, the analysis included 12 randomized, controlled trials of 2,060 patients. Pharmacist care was associated with a significant reduction in all-cause hospitalization (OR = 0.71; 95% CI, 0.54–0.94) and heart failure hospitalization (OR = 0.69; 95% CI, 0.51–0.94). The impact on mortality was not statistically significant (OR = 0.84; 95% CI, 0.61–1.15).

The output from a meta-analysis is usually expressed as an OR, not an RR. Although the two are often similar and interpreted in a similar fashion, they are distinctly different (Table 6). The odds of an event are calculated by dividing the number of subjects with an event by the number of subjects without that event. The OR, then, is the odds of an event occurring in the intervention group compared with (i.e., divided by) the odds of the event occurring in the control group (contrast this with RR, which is the event rate in the intervention group divided by the event rate in the control group).

Table 6. Calculating an RR/OR

		Disease	
		Yes	No
Exposure	Yes	a	b
	No	c	d

OR = odds ratio; RR = relative ratio.
$RR = (a/(a + b))/(c/(c + d))$; $OR = (a/c)/(b/d)$.

Results of systematic reviews are commonly reported as plots that include the center (i.e., point estimate) and CI of the difference reported from each individual study, together with the center and CI of the results of the meta-analysis (Figure 11).

In the 2008 review, the OR for the impact of pharmacist care on heart failure hospitalization was 0.69. This is often interpreted as similar to the RR of 0.69 (i.e., an RRR of 31%; or, pharmacist care reduces heart failure hospitalization by 31%). Although not technically correct, this is close enough for most purposes.

The main advantage of systematic reviews is that they provide the highest level of evidence by considering all available data. They can also provide a good quantitative estimate of the effect size of a given treatment. However, the conduct of systematic reviews contains several pitfalls. First, garbage in leads to garbage out: if there are no good-quality trials on which to conduct a meta-analysis, no amount of analytic skill will produce a valid result. In addition, an investigator can never be completely certain whether the constituent trials were randomized properly.

Publication bias is another potentially important source of error; this occurs when only positive studies are published, and negative or

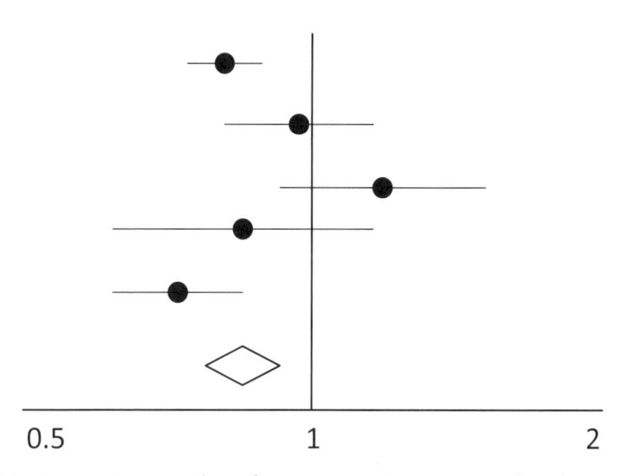

Figure 11. Reporting results of a systematic review. The diamond is the composite OR and CI, whereas each dark circle and line represents the OR and CI, respectively, of the constituent trials included in the meta-analysis. CI = confidence interval; OR = odds ratio.

indeterminant trials are not pursued to publication. Publication bias is why trial registries are important; such registries hold authors accountable to provide results, whatever they may be.

Another limitation of systematic reviews is the heterogeneity between studies (i.e., "combining apples and oranges"). This heterogeneity may be clinical (different conditions or stages of disease), methodological (different research designs), or statistical (different treatment effects). Although statistical tests for heterogeneity are available, some clinical judgment is required. If heterogeneity is present, then combining data in a meta-analysis could be misleading and should not be done.

SELECTED BIBLIOGRAPHY

The CAPS Investigators. The Cardiac Arrhythmia Pilot Study. Am J Cardiol 1986;57:91–5.

Fleming TR, DeMets DL. Surrogate end points in clinical trials: are we being misled? Ann Intern Med 1996;125:605–13.

Guyatt GH, Sackett DL, Cook DJ; for the Evidence-Based Medicine Working Group. Users' guides to the medical literature. II. How to use an article about therapy or prevention. A. Are the results of the study valid? JAMA 1993;270:2698–701.

Guyatt GH, Sackett DL, Cook DJ; for the Evidence-Based Medicine Working Group. Users' guides to the medical literature. II. How to use an article about therapy or prevention. B. What were the results and will they help me in caring for my patients? JAMA 1994;271:59–63.

The Heart Outcomes Prevention Evaluation Study Investigators. Effects of an angiotensin-converting enzyme inhibitor, ramipril, on cardiovascular events in high-risk patients. N Engl J Med 2000;342:145–53.

Kaul S, Diamond GA. Good enough: a primer on the analysis and interpretation of noninferiority trials. Ann Intern Med 2006;145:62–9.

Koshman SL, Charrois TL, Simpson SH, McAlister FA, Tsuyuki RT. Pharmacist care of patients with heart failure. A systematic review of randomized trials. Arch Intern Med 2008;168:687–94.

MERIT-HF Study Group. Effect of metoprolol CR/XL in chronic heart failure: Metoprolol CR/XL Randomised Intervention Trial in Congestive Heart Failure (MERIT-HF). Lancet 1999;353:2001–7.

The ONTARGET Investigators. Telmisartan, ramipril, or both in patients at high risk for vascular events. N Engl J Med 2008;358:1547–59.

Oxman AD, Cook DJ, Guyatt GH; for the Evidence-Based Medicine Working Group. Users' guides to the medical literature. VI. How to use and overview. JAMA 1994;272:1367–71.

Preliminary report: effect of encainide and flecainide on mortality in a randomized trial of arrhythmia suppression after myocardial infarction. The Cardiac Arrhythmia Suppression Trial (CAST) Investigators. N Engl J Med 1989;321:406–12.

Scandinavian Simvastatin Survival Study Group. Randomised trial of cholesterol lowering in 4444 patients with coronary heart disease: the Scandinavian Simvastatin Survival Study (4S). Lancet 1994;344:1383–9.

Truett J, Cornfield J, Kannel W. A multivariate analysis of the risk of coronary heart disease in Framingham. J Chronic Dis 1967;20:511–24.

SELF-ASSESSMENT QUESTIONS

1. A systematic review evaluated the effect of bisphosphonates on skeletal metastasis in patients with cancer. A meta-analysis that included placebo-controlled trials of at least 6 months' duration showed a combined odds ratio (OR) for the use of radiation therapy of 0.67 (95% confidence interval [CI], 0.57–0.79). For spinal cord compression, bisphosphonates compared with placebo resulted in a combined OR of 0.71 (95% CI, 0.47–1.08). Which one of the following is the best interpretation of these results?

   A. Bisphosphonates favorably affect the use of radiation therapy and the occurrence of spinal cord compression in patients with cancer with skeletal metastasis.

   B. Bisphosphonates affect neither the use of radiation therapy nor the occurrence of spinal cord compression in patients with cancer with skeletal metastasis.

   C. Bisphosphonates favorably affect the use of radiation therapy but not the occurrence of spinal cord compression in patients with cancer with skeletal metastasis.

   D. Bisphosphonates do not affect the use of radiation therapy but do favorably affect the occurrence of spinal cord compression in patients with cancer with skeletal metastasis.

2. A systematic review reports that its meta-analysis includes studies of the following sizes: two studies with sample sizes between 200 and 400 patients; four studies with sample sizes between 401 and 1,000 patients; one study with a sample size of 1,100 patients; and one study with a sample size of 5,200 patients. In reviewing the results for use in practice, which one of the following is the most important type of analysis to seek out?

   A. A regression analysis of variables that might increase the risk of the outcome of interest

   B. A sensitivity analysis and tests for heterogeneity

   C. A calculation of the hazard function for the total number of patients

   D. A Cox regression analysis

3. A physician in your hospital wants to start all of his patients admitted to the intensive care unit on a proton pump inhibitor. He states that a recent study of hospitalized patients found that proton pump inhibitors improve survival when used for stress ulcer prophylaxis. You realize this will increase the pharmacy's budget considerably and want to thoroughly evaluate the evidence. Which one of the following statements is true regarding your evaluation of the article to which the physician refers?

   A. The inclusion and exclusion criteria should be evaluated to determine whether the patient population is similar to your patient population before using the study results to make a decision.
   B. If a significant difference is found between the two groups, then the possibility of a type II error must be considered.
   C. It is not necessary to review the competency of the authors by reviewing their affiliations and credentials.
   D. The methods used to blind the study are not important because the outcome was objective.

4. As an oncology pharmacist at a large tertiary care center, you determine that one of your patients is experiencing Stevens-Johnson syndrome. The patient recently was initiated on Curemesis, an antiemetic you often recommend for your patients. After reviewing pertinent tertiary references, you learn that Curemesis has been associated with Stevens-Johnson syndrome in less than 1% of patients. You research this further, trying to determine the symptoms of the reaction and the actions of other clinicians in treating patients experiencing Stevens-Johnson syndrome caused by Curemesis. Which one of the following types of clinical trial design is the best source to find the information you need?

   A. Randomized, controlled trial
   B. Case report or case series
   C. Meta-analysis
   D. Case-control study

5. A new drug is indicated for type 2 diabetes mellitus by the U.S. Food and Drug Administration (FDA) on the basis of a single randomized,

controlled trial. The patients you see in the clinic primarily are men and women older than 50 years with concomitant cardiovascular disease. You also have many patients who have failed to benefit from therapy with sulfonylureas and metformin. Which one of the following will most likely affect your ability to extrapolate the results of this trial to the patients in your clinic?

A. The inclusion criteria of this study were men and women older than 18 years with type 2 diabetes mellitus.
B. The exclusion criteria were a history of cardiovascular disease, liver disease, and concomitant use of insulin.
C. The mean age of the patient population studied was 52.5 years, and 60% of the population was male.
D. Most of the patients' diseases included in the study had been uncontrolled on metformin.

6. You are a member of the Pharmacy and Therapeutics Committee at your hospital. The hospital is considering adding levalbuterol to the formulary. Most comparative studies with levalbuterol show its similarity to albuterol in efficacy, with a small difference in favor of levalbuterol in the effect on heart rate. One comparative study found levalbuterol to be associated with a shorter length of hospital stay and decreased hospital costs. However, this study was a retrospective chart review. Which one of the following is true regarding this literature?

   A. The retrospective design of this study makes the results subject to recall bias.
   B. Levalbuterol should be added to the formulary because of the potential cost-savings, as shown in the retrospective study.
   C. Retrospective studies are the strongest means by which to determine cause and effect.
   D. Levalbuterol should be added to the formulary because of its improved safety in its effect on heart rate, as shown in comparative trials.

7. Parknia is a new drug with Parkinson disease as its labeled use. The drug's labeling is based on the results of three randomized, placebo-controlled trials. The physicians in your clinic want to start prescribing the drug, but they want to know how it compares with the drugs

they currently use. You conduct a MEDLINE search to see whether any newer trials have been published on the drug; you find two randomized, controlled trials that compare Parknia with drugs in current use. Both studies conclude that Parknia is similar in efficacy to the other drugs. Which one of the following is true regarding the evaluation of these studies?

A. The doses of the comparative drugs should be the same as those used in clinical practice.
B. The possibility of type I error should be considered.
C. Because the two studies are not placebo controlled, the results are invalid.
D. If the study sponsor were disclosed, its role would not be important to evaluate.

8. One of your colleagues has noticed that more pharmacy technicians have latex allergy. You wish to determine the prevalence of latex allergy in a population of pharmacy technicians. Which one of the following clinical trial designs is best suited for this purpose?

A. Randomized, controlled, clinical trial
B. Case-control study
C. Cohort study
D. Cross-sectional study

9. A new drug, Superstatin, is indicated for treating hyperlipidemia; however, the FDA is concerned that this drug may cause liver toxicity. There were a few reports of liver toxicity before the drug came to the market; however, not enough reports existed to conclude an increased risk of liver toxicity. Which one of the following study designs is best suited to identify whether an increased risk of liver toxicity exists with Superstatin?

A. Randomized, controlled, clinical trial
B. Case-control study
C. Prospective cohort study
D. Cross-sectional study

10. Some reports have stated that low-dose aspirin may decrease the risk of certain types of cancer. You have a large population of patients

with colon cancer at your hospital and want to conduct a study to determine whether the use of low-dose aspirin is associated with a reduced risk of developing colon cancer. Which one of the following clinical trial designs is best suited for this purpose?

A. Randomized, controlled, clinical trial
B. Case-control study
C. Prospective cohort study
D. Cross-sectional study

11. You are appraising a meta-analysis of studies measuring the association between pediatric vaccinations and the development of autism. You are trying to determine whether you need to warn the parents of your pediatric patients of the risk. Which one of the following is true regarding the evaluation and application of the results of this meta-analysis?

    A. If the investigators searched MEDLINE and the reference lists of all identified articles, the study would be especially susceptible to the effects of publication bias.
    B. The conclusions of this meta-analysis are stronger than those in a randomized, controlled trial and should be used to guide clinical practice.
    C. One disadvantage of meta-analyses is that they increase the chance for type II error.
    D. One advantage of meta-analyses is that they increase the heterogeneity of the studies being analyzed, which improves the statistical analysis of the results.

12. You are evaluating a cohort study on the relationship between alcohol consumption and the development of certain types of cancer. The investigators included 300 patients who drank more than 12 oz/day of alcohol and 300 patients who did not drink alcohol. The risk of developing cancer was 20-fold higher in the patients who drank alcohol than in those who did not. The authors also found that cigarette smoking was more common in the alcohol-drinking group. Which one of the following limitations or biases is most likely to have occurred in this study?

A. Confounding
B. Decreased external validity
C. Increased internal validity
D. Prevalence bias

13. You are analyzing the results of a case-control study of the effects of ephedra use on weight loss. Patients who had lost weight during the past 6 months were identified. The subjects were the patients who had used ephedra to lose weight, whereas the controls were the patients who had not used ephedra to lose weight. The subjects were questioned regarding the amount of ephedra they had taken, other methods used to lose weight, and the total amount of weight lost while taking ephedra. The control group also was questioned about methods used to lose weight and how much their weight had fluctuated during the same period. Which one of the following is your biggest concern with the methods of this study?

    A. External validity
    B. Blinding
    C. Randomization
    D. Recall bias

Questions 14 and 15 pertain to the following case.
To assess whether there is a difference in the occurrence of clinical response between patients receiving antiepileptic drug A versus antiepileptic drug B, investigators conduct a secondary data analysis. They analyze all patients to whom drug A and drug B are prescribed, and the investigators have 60 days of data after patients receive the drug during year 2003 in a posttest-only with control group design. The primary outcome is the average number of seizures per week during the 60-day period, as measured by patient self-report through a diary. Seizure history has been accurately captured in the diaries and database.

14. Which one of the following is the most important threat to internal validity in this study?

    A. Repeated testing
    B. Selection bias
    C. Experimental mortality
    D. Instrumentation

15. Suppose patients taking drug A had 15% fewer seizures per week than patients taking drug B. With no change in design or analysis plan, which one of the following valid statements can be made regarding the effectiveness of these drugs?

    A. Patients taking drug A had 15% fewer seizures per week than patients taking drug B during the study period.
    B. Drug A is more effective at preventing seizures than drug B ($p<0.005$).
    C. Drug A is more effective at preventing seizures than drug B ($p>0.005$).
    D. The relative frequency of seizures between the study groups cannot be accurately assessed.

16. You are reading about the trial of a new renin inhibitor in patients with heart failure and the drug's effect on B-type natriuretic peptide (BNP) concentrations. Which one of the following is the best statistical test for the trial?

    A. The t-test, because the data are dichotomous
    B. The t-test, because the data are continuous
    C. The chi-square test, because the data are dichotomous
    D. The chi-square test, because the data are continuous

17. You wonder why the investigators in the previous question did not report their data the way you would interpret the results clinically (i.e., concentrations less than 100 pg/mL = not in heart failure; 100–499 pg/mL = possible heart failure; and greater than 500 pg/mL = heart failure). Which one of the following best characterizes why investigators should report actual BNP values?

    A. Categorizing BNP values requires the use of arbitrary cutoff values.
    B. Reporting continuous BNP values is essential for surrogate markers of disease.
    C. Reporting continuous BNP values is more powerful than categorizing them.
    D. It is unlikely that any treatment would affect BNP concentrations enough to change a patient's category.

Questions 18–20 pertain to the following case.

A 3-year randomized, controlled trial is conducted of a new antithrombotic agent in patients with atrial fibrillation. Researchers find the rate of stroke in the treatment group to be 15%, compared with a rate of 21% in the control group.

18. What is the absolute risk reduction (ARR) in this trial?
    A. 6%
    B. 71%
    C. 29%
    D. −6%

19. What is the relative risk reduction (RRR) in this trial?
    A. 6%
    B. 71%
    C. 29%
    D. 17%

20. What is the number needed to treat (NNT) in this trial?
    A. 6
    B. 71
    C. 29
    D. 17

21. A recently published trial showed an RRR of 40% with a 95% confidence interval (CI) of −0.99% to 55.5%. Which one of the following is the best way to describe these results?
    A. The CI is wide.
    B. The finding is statistically significant, with a p value less than 0.05.
    C. The investigators have not excluded a potentially important 55% increase in strokes.
    D. The results are not statistically significant.

22. A group of researchers conducted a randomized, controlled trial to determine the effect of a new antihypertensive on blood pressure. The original sample size calculation, designed to attain 80% power and an α of 0.05, required a sample of 563 subjects per treatment

arm. However, the researchers ran out of funding when they had 400 patients per group. The analysis was still conducted with 800 subjects. What is the result of this decision?

A. The α decreases.
B. The α increases.
C. The power decreases.
D. The power increases.

23. A randomized multicenter trial evaluates the impact of a new drug-eluting stent on repeat target vessel revascularization (a dichotomous variable). Which one of the following would most likely cause major losses in study power?

A. Underestimation of the efficacy of the new stent
B. Protocol violations in the conduct of the study
C. Overestimation of the event rate in the control group
D. Inability to blind the intervention

Questions 24–26 pertain to the following case.
A randomized, controlled trial compares new drug A with placebo with the primary outcome of death. Several patient demographic and comorbidity variables were collected. At baseline, the age of the treatment group was much higher than that of the placebo group, and patients in the placebo group had more instances of heart failure.

24. Researchers decided to use logistic regression for analysis of the primary outcome. The type (continuous or dichotomous) of which one of the following variables best explains this decision?

A. Outcome variable
B. Outcome variable and independent variables
C. Independent variable only
D. Outcome variable or independent variables

25. A multiple logistic model was used in this trial. Which one of the following best determines whether researchers should use simple or multiple regression?

A. Whether the outcome is continuous or dichotomous
B. Whether the dependent variable is single or multiple

C. Whether the independent variable is single or multiple
D. The clinical characteristics chosen to enter in the model

26. The use of a multiple logistic model in this randomized, controlled trial will most probably allow the investigators to accomplish which one of the following?
    A. Find the predictors of mortality
    B. Find the treatment effect
    C. Adjust for baseline discrepancies
    D. Find variables that affect treatment

Questions 27 and 28 pertain to the following case.
A new drug to be used in the management of heart failure is being tested in a randomized, controlled trial. The researchers report the primary outcome of death using a survival curve.

27. Which one of the following insights is most likely gained from the survival analysis?
    A. The temporal course of treatment effects
    B. The temporal course of disease processes
    C. The tracking of only "bad" events
    D. The same follow-up duration for all patients

28. A patient in the treatment group of this randomized, controlled trial is lost to follow-up. Which of the following is the best way to handle this in the survival analysis?
    A. The patient is excluded from the analysis because of incomplete information.
    B. The time until the loss to follow-up is still used.
    C. The person is thought to have an event rather than no event.
    D. The outcome of another patient with similar demographics is used to substitute the outcome values.

Questions 29 and 30 pertain to the following case.
You are on the advisory board of a pharmaceutical company that has developed a new anticoagulant for stroke prevention in atrial fibrillation.

As you consider the study design for a phase III study, a colleague asks whether a non-inferiority trial would be desirable for this new agent.

29. Which one of the following best describes the appropriate use of non-inferiority trials?
    A. When the standard treatment is of unknown or unclear efficacy
    B. When there is no established treatment
    C. When it is not possible to obtain sufficient funding for a superiority trial
    D. When a placebo-controlled trial is not ethical or possible

30. A non-inferiority trial compares the new anticoagulant (treatment A) with warfarin (standard treatment B). Which one of the following best describes what the researchers set out to prove (the alternative hypothesis)?
    A. Treatment A is better than treatment B by a factor of Δ.
    B. Treatment A is worse than treatment B by a factor of Δ.
    C. Treatment A is not worse than treatment B by a factor of Δ.
    D. Treatment A is the same as treatment B.

31. You are a member of a national Clinical Practice Guidelines Committee for acute coronary syndromes. As such, you are surrounded by many opinionated subspecialists who have conducted research in this field. The guidelines committee has agreed that the document prepared should be evidence based. You suggest that greater weight should be placed on the results of high-quality systematic reviews of randomized trials. Which one of the following best describes why systematic reviews are considered the highest level of evidence?
    A. They reduce bias.
    B. They include a large number of studies and patients.
    C. They combine data across studies to provide a quantitative estimate of the real effects of treatment.
    D. They summarize the current medical literature.

32. Which one of the following is the most important pitfall in the conduct and/or interpretation of systematic reviews?

A. The quality of available trials
B. Analytic techniques for combining data
C. Understanding of the application of odds ratios to clinical practice
D. Validity of combining data across different studies

33. Which one of the following best describes why surrogate outcomes may be misleading?
    A. They are continuous variables.
    B. The studies are generally smaller.
    C. They are based on pathophysiologic processes.
    D. They may not represent the clinical outcome of interest.

34. Many cardiovascular trials now use composite outcomes such as heart failure hospitalization or death. Which one of the following best describes why composite outcomes are useful?
    A. They can provide data on adverse effects in large populations.
    B. They add together outcome events to form a single end point for a study.
    C. They reduce the complexity of describing treatment effects.
    D. They can the reduce sample size requirement.

35. Which one of the following describes the most important limitation of composite outcomes?
    A. They allow the counting of many events in the same patient.
    B. They assume that all outcomes are equally important.
    C. There is insufficient power to detect treatment effects on the individual components of the composite outcome.
    D. They are unable to classify all outcomes precisely.

Pharmacoepidemiology

INTRODUCTION

Epidemiology is the rigorous and scientific study of the distribution and determinants of disease (and other outcomes) in humans. The goals of this effort include applying this information to improve the structure, process, and outcomes of health care and developing a commonsense health policy that is rooted in evidence. Pharmacoepidemiology uses techniques of epidemiology to study the use and effects of drugs.

Epidemiologic research includes descriptive studies, which estimate the natural history of disease and exposure, and inferential studies, which provide measures of association between exposure and outcomes. Incidence and prevalence rates are examples of descriptive epidemiologic data, whereas inferential estimates include measures such as ORs, RRs or rate ratios, and hazard ratios. Data used to conduct these studies come from a variety of sources, including administrative databases (i.e., Centers for Medicare and Medicaid Services and insurance data), clinical study data, patient registries (i.e., the Surveillance Epidemiology and End Results registry), and other large studies.

As pharmacy practice evolves, incorporating epidemiologic skills into everyday practice makes increasing sense. These skills help pharmacists contribute to outstanding evidence-based health care, communicate with patients and other health care providers using a common language, and answer research questions about the role of drugs in caring for people. The following is a review of selected epidemiologic concepts and tenets of study design useful to clinical pharmacists when interpreting the literature, as well as a discussion of pharmacoepidemiologic issues expected to contribute to clinical practice.

CAUSATION

The goal of scientific research is to observe, describe, and, ultimately, relate observations to some cause. Teasing out the different threads of causation is difficult, however, and there is no definitive list or set of methods that allows us to determine whether an association is causal. One set of often-used causality criteria, often called the Bradford-Hill criteria, consists of strength, consistency, specificity, temporality, biologic gradient, plausibility, coherence, experimental evidence, and analogy. For example, the risk of lung cancer in smokers is about 10 times that in nonsmokers. In general, the idea of strength is used to convey a sense of risk. If the risk of disease in individuals exposed to a factor is not much higher than in unexposed individuals, the factor is a "weak" cause. The second criterion, consistency, applies when the same general association is observed for different groups of people in different circumstances. For a suggested cause to be specific, it must lead to just one effect, a difficult requirement to fulfill. Temporality requires that the exposure come before the disease; in the smoking example, it makes sense that, to develop a smoking-related cancer, the person must first have smoked. A biologic gradient is present when there is a positive dose-response relationship: as dose increases (i.e., smoking exposure), the chance of developing disease also increases. A plausible hypothesis is one that makes sense, though this criterion is relatively subjective and depends on other factors. For example, if the mechanism of action of a drug is not well understood, plausibility may not be obvious. Coherence is similar to plausibility; this criterion requires that the explanation suggested for an observation not conflict with what is known about the pathophysiology of the disease, such as indicators of damage to bronchial tissue caused by exposure to cigarette smoke. The criterion of experimental evidence has been suggested to refer to data indicating that the removal of a risk factor had a beneficial effect. Finally, this analogy refers to the ability to conceive of more complicated theories of the relationship between the suggested cause and the disease being studied.

This set of conditions is commonly used, but important problems exist with each criterion, as described by Kenneth Rothman in his introductory biostatistics text. For example, the strength of a cause is relative, meaning that it depends on whether other causes are known. Smoking is considered a stronger cause of lung cancer compared with exposure to radon gas because more cases are ascribed to smoking than to radon gas exposure. If smoking is ever eliminated, and the cases of lung cancer related with it, other causes will appear to be stronger causes of lung cancer. Consistency is a relatively poor indicator of a relationship between exposure and outcome because causality is a function of the exposure, the individual's genetic makeup, and the person's environment. A cause may have many effects, and many exposures may result in the same outcomes; thus, the relationship of cause to effect and vice versa is often inconsistent. To further complicate matters, records are notoriously incomplete and human memory is fallible, so even if an exposure perfectly predicts an outcome, establishing a timeline is often difficult. The problem is enhanced when exposure occurs a long time before the outcome or when exposure is intermittent. Plausibility requires the observer to make a subjective judgment, which may or may not correspond to other observers' point of view. Coherence and analogy are inherently unclear, and they are closely related to other criteria. Experimental evidence is not always available, or its generalizability may be limited.

Other challenges exist in considering relationships between exposures and outcomes, such as whether a risk factor is necessary or sufficient to result in the outcome of interest. Exposure to a necessary factor is just that: without a specific exposure, the outcome does not occur. Unless a person is exposed to the varicella zoster virus, chickenpox cannot occur. A sufficient exposure, however, is enough to cause disease. For example, exposure to tobacco smoke is a primary risk factor for developing chronic obstructive pulmonary disease. Although smoking tobacco (or similar exposure) accounts for most of the risk of developing chronic bronchitis or emphysema, not everyone who smokes will develop these conditions. Similarly, smoking cigarettes is a well-known risk factor for developing lung cancer, and most people who develop lung cancer have smoked at some point in their lives. Even though smoking exposure is an important contributor to these conditions, some individuals who do not smoke will develop these outcomes. Smoking is neither necessary nor sufficient to produce these results. Because of these and other limitations, some authors advocate assessing causation from a refutationist approach in which a cause can be ruled out, but never completely confirmed.

STUDY DESIGN

Randomized, controlled studies are considered the gold standard for comparing health care interventions, but not all randomized studies are the same. End points between studies often differ; poorly designed studies may result in errors that have direct clinical, economic, and human costs; and not all scientific questions can be studied using a randomized design. In addition, many clinical decisions (and everyday choices) are made without the benefit of clinical study data. Although well-designed randomized studies are useful tools, clinicians should be familiar with study design issues for each major type of study used in clinical practice, particularly randomized, controlled, cohort, and case-control studies.

Randomized, Controlled Studies

In clinical studies, the experiences of members of an intervention group and a control group are compared. When the individuals taking part in a study are assigned to a treatment group without respect to patient characteristics, the result is a randomized, controlled, clinical study.

The main advantages of randomized, controlled studies are that, when performed correctly, randomization prevents biases caused by imbalanced allocation, results in similar groups, and helps ensure that statistical testing results are valid. Randomization also prevents confounding, including channeling bias. Although randomized clinical trials are statistically the gold standard, several inherent problems limit the application of their results to clinical practice. For example, randomized, controlled trials commonly study only the effect of one drug, whereas our patients receive many medications and undergo many procedures that may influence the outcome. Other major disadvantages of these studies are that they are time-consuming, and expensive to conduct; also, not every scientific question can be evaluated using this design. For example, study participants cannot be randomized to be exposed to interventions known to be harmful, such as smoking.

Because resources (e.g., money and time) are limited, it is important to consider whether a randomized design is the most appropriate study design to test a hypothesis. One part of the decision-making process involves whether there is equipoise, or uncertainty, about the risks and benefits of the interventions being studied. When there is equipoise, a randomized study may be indicated. If there is not equipoise, however, a different approach to studying the question is likely a better choice.

Cohort Studies

In cohort studies, the experience of individuals in different groups whose members share some trait in common is examined. Examples of cohorts include students in a graduate program at a particular school, children monitored at a specific medical practice, individuals with a specific genetic trait, and adults admitted to a hospital.

In this type of study, cohort members are observed over time, and the experiences of the members of one cohort are compared with those of other cohorts to describe outcomes among people whose exposure status is known and who are at risk of experiencing the event. Because cohort members must be at risk of experiencing the event, they also must be known not to have had the outcome of interest at the beginning of the study. For example, if we are interested in assessing all-cause mortality in a cohort of individuals with diabetes, study participants must have definitively been given a diagnosis of diabetes; must be alive (i.e., at risk of the event—all-cause mortality); and, because they are at risk, must have not yet died. In pharmacoepidemiologic research in which there is interest in non-death outcomes, it is sometimes difficult to clearly establish that the person did not have the outcome by the start of follow-up. The issue is that unrelated events will be attributed to the exposure, introducing bias, particularly if the rate of the event is different between groups.

To reduce the potential for this type of bias, some researchers require that study participants be members of the cohort for some minimal amount of time before the start of exposure. This amount of time is chosen on the basis of evidence and expert opinion about how long a subclinical incident event will take to appear. Although cohort members must be alive and must not have had the event when the study starts, this requirement is potentially problematic if the event can occur more than once. The difference is that a person who experiences an outcome that can happen only once stops contributing time to the analysis. Alternately, if an event can occur more than once, an individual may remain in the cohort as long as he or she continues to remain at risk of the outcome.

Cohort data can be used to estimate the risk or incidence rate ratio. Risks are estimated by dividing the number of new events by the number of individuals being observed for the exposed and unexposed cohorts. In addition, cohorts may be open or closed. In an open cohort, people can enter or exit the group as the study progresses. The membership of a closed cohort is fixed when follow-up begins. No additions are allowed; however, the size of the cohort is expected to shrink during the study, as study participants are lost to follow-up, as they die, or as they experience the event of interest. It is more difficult to estimate risk in an open cohort study because of possible changes in the sample size, though it may also be reasonable to choose an observation period that is shared by enough cohort members to address this issue. An alternative in either type of cohort study—and the logical approach in open cohort studies—is to add the person-time and then divide the number of events by the total person-time contributed by cohort members.

Case-Control Studies

In addition to cohort studies, another primary type of epidemiologic study is the case-control study, in which individuals with a particular outcome (the cases) are compared with individuals without that outcome (the controls). As in a cohort study, the same cases are identified as exposed or unexposed. Members of the control group are chosen from the underlying population the cases came from, and then they are also identified as exposed or unexposed. Because most people at risk of developing a particular outcome experience that event, the case-control study seeks to accomplish the same goal as a cohort study, but more efficiently. The case-control design addresses the largest disadvantage to conducting a cohort study: the time and expense needed to collect data on the cohort members. Of note, controls in a case-control study are selected independently of their exposure status, unlike in a cohort study. When this goal is met and when the members of the control group are chosen using incidence density or risk-set sampling, in which controls are selected from among individuals at risk at the time of the event, the control group's experience represents the person-time distribution of exposure, and the OR estimates from a case-control study represent the incidence rate ratio.

Nested case-control studies are designed to take place within a cohort, but it is straightforward to consider all case-control studies as nested because they are all contained within some source population. Because the main advantage of the case-control design over a cohort study is efficiency, there is no particular advantage to the nested variant when all the data are readily available, as in many database analyses.

Comparing Cohort and Case-Control Studies

The main difference between cohort and case-control studies involves exposure status. In a cohort study, the exposure status of the cohort is known from the outset, although it is possible for exposure status to change over time, as in time on drug analyses. In a case-control study, exposure status may be known, whereas controls are selected independently of exposure status. Each study type has particular features as well as advantages and disadvantages. In general, cohort studies completely describe the source population at risk; as a result, the incidence rates, risks, differences, and ratios for each cohort can be estimated directly. In addition, because exposure status is known from the outset of follow-up, cohort studies can be used to assess the relationship of the exposure to many outcomes. Cohort studies are generally expensive and inefficient in studying rare events. In contrast, although risks or incidence rates can be estimated from case-control studies conducted within fully specified cohorts, if the case-control study is conducted within a virtual cohort, the investigator is usually limited to estimating risk or rate ratios. Table 6 illustrates the difference between calculating an OR and an RR. Finally, although case-control studies are typically considered more efficient than cohort studies, the use of computer time and the verification that exposure and person-time have been correctly classified require a great deal of resources.

Other Types of Studies

The case-cohort and case-crossover are two other types of study designs used in epidemiologic research. In the case-cohort study, a sample of the underlying population is chosen for the control group. As a result, each control in a case-cohort study represents a portion of the total number of individuals in the source population rather than a fraction of the person-time. Because members of the control group are chosen from all cohort members rather than from person-time experience, the OR from a case-cohort study represents the risk ratio rather than the incidence rate ratio.

In crossover and case-crossover studies, study participants serve as their own control. Unlike a crossover study, in which each person receives each treatment in random order, all individuals in a case-crossover study have had the event of interest, and the control represents the time before the event, not a separate group of people. This type of study is useful to assess sudden events, including automobile accidents and MIs, as well as other events in which the exposure varies in the same person.

Retrospective and Prospective Designs

Retrospective and prospective refer to the conduct of the study rather than the actual design. A prospective case-control study uses exposure data collected before the outcome occurred, whereas a retrospective case-control study uses exposure data collected after the development of disease. Thus, equating retrospective and prospective with case-control and cohort, respectively, is incorrect. Similarly, assuming that the study is appropriately designed and carried out, the idea that retrospective studies are somehow less valid or useful than prospective studies is erroneous. The timing of the study has nothing to do with the validity of the resulting estimates. Furthermore, discounting results from well-designed, carefully conducted studies—whether they are retrospective, prospective, case-control, or cohort—because of this misguided belief could be harmful if it causes us to ignore information that provides insight into scientific questions.

MEASUREMENTS OF OCCURRENCE

The vocabulary of pharmacoepidemiology is designed to put different estimates of occurrence and association into perspective. As a result, it is useful to spend some time reviewing a few concepts.

Risk refers to the chance that an individual or members of a group will experience a particular outcome over some time interval. Estimates of risk are generally more meaningful in the context of groups, where the average risk over time refers to the incidence proportion. Without knowing the time interval, risk and incidence proportion are not easily interpreted. For example, everyone will eventually die, but of interest is the rate of the event. The risk of death is one apiece for everyone, or 100%, but the risk of death in specified intervals (age 50–55, during the next year, or per passenger-mile) varies and serves as a useful basis for comparisons.

Incidence is a measure of occurrence among individuals who are at risk of the event. The incidence proportion is the percentage of the group who had the outcome, whereas the incidence rate reflects the number of individuals who have the outcome of interest over some interval. The primary difference between these two measures is that the incidence proportion is calculated as a simple percentage of the group being observed, and the incidence rate incorporates the time observed. As a result, if a group of 10,000 individuals were observed for a full year and, by the end of the year, 30 individuals had experienced a potentially recurring outcome such as the common cold, the incidence proportion would be expressed as a probability (30/10,000) without units. In contrast, the incidence rate would be expressed as 30 individuals per 10,000 person-years, where the denominator of the rate is the sum of time at risk of the outcome for each person in the group being studied. If the outcome of interest were an event that could not recur, it would be necessary to remove the person-time contributed by those individuals once the event had occurred. If the events occurred evenly throughout the year, the lost person-time would be 30/2, or 15, person-years. The denominator would be reduced to 9,985 (10,000 − 15) person-years, and the incidence rate would be 30.05 per 10,000 person-years.

In contrast to incidence, the prevalence proportion, or prevalence, refers to status. Prevalence is an estimate of the number of individuals who have or have had the event of interest. For example, in a study of a particular adverse drug event, prevalence is the percentage of individuals who experience the reaction at some specified time.

Effect Measures

Incidence and prevalence are the main epidemiologic measures of occurrence, but they are not the only estimates used. Other measures used to compare what happens in exposed individuals relative to unexposed individuals include the OR, RR, risk difference, incidence rate difference, and attributable fraction. These measures are described in Table 7. In general, each of these measures of effect compares the experience among exposed individuals with that among unexposed individuals.

Risk expresses the probability that an individual will develop a given condition. Because we usually refer to populations or at least large groups of people in pharmacoepidemiologic research, risk is often phrased as the proportion of the group that will eventually become affected. The risk of an event can be calculated as the number of individuals who develop a disease over a period divided by the number of people who are observed for that interval. In estimating risk, everyone in the denominator group should be observed for the entire observation period. The term *incidence proportion* is often used to describe the average experience of a group.

Although the concept of risk is easy to define, interpreting risk estimates is complex because, by itself, the proportion of people who develop a condition or disease tells us nothing specific about the observation period. For example, imagine that we are interested in how many people who use a nonsteroidal anti-inflammatory drug will have a cardiovascular event. Unless the observation period is clearly defined, the resulting estimate will not be interpretable.

Table 7. Commonly Used Epidemiologic Measures

Measure	Equation
Risk	$\frac{\text{number of people developing disease or event during a period}}{\text{number of people observed for the period}}$
Incidence rate	$\frac{\text{number of incident cases}}{\text{amount of at-risk experience}}$
Prevalence	$\frac{\text{number of prevalent cases}}{\text{size of population}}$
Risk difference	$\text{incidence}_{\text{exposed}} - \text{incidence}_{\text{unexposed}}$
Relative risk (RR, risk ratio)	$\frac{Incidence_{\text{exposed}}}{Incidence_{\text{unexposed}}}$
Odds ratio	$\frac{ad}{bc}$ (from a 2 × 2 table)
Attributable fraction (percent)	$\frac{\text{RR-1}}{\text{RR}} \times 100\%$

The incidence rate refers to the number of people who develop a condition divided by the total time experienced for the people observed and reflects the risk during the next very small period. Although calculating the denominator of an incidence rate is reasonably straightforward—the time that each person in the group is observed is added—this effort depends on whether the event in question can occur more than once. Some events, such as death, can happen only once. With others, such as adverse drug reactions, it may be possible for a person to have the event more than once, but it is often unlikely, because the individual will likely avoid the drug in question. Although incidence refers to the frequency of events from a given pool of person-time, prevalence refers to the proportion of people in a group who have (or have had) the disease, condition, or event of interest at some particular time.

The risk difference is the difference in incidence proportion or risk between members of these groups. If the incidence proportion is replaced with the incidence rate and the calculation is repeated, the result is, not surprisingly, the incidence rate difference. The advantage of this latter estimate over the risk difference is that with the incidence rate, there is some understanding of the role of time.

The risk ratio, sometimes called the RR (relative risk), is the risk of a particular event in exposed individuals divided by the risk of that same event in unexposed individuals. Subtracting 1 from this estimate results in the relative effect, but the RR is often used as the measure of effect without subtracting 1. Because risk cannot be negative, neither can the RR. An RR of 1 indicates that the risk in exposed and unexposed individuals is the same. As a result, an RR of 2 indicates a 100% increase in risk among exposed individuals. Similarly, we can interpret an RR of 2.5 as being 150% greater than that of 1. The incidence rate ratio is determined by dividing the incidence rate in exposed individuals by that in unexposed individuals; subtracting the incidence rate in unexposed individuals from that in exposed individuals; and dividing by the incidence in exposed individuals. This provides an estimate of the percentage of events that result from exposure, also known as the attributable fraction.

It is also useful to examine risk ratios and rate ratios more closely. The difference is important because although the risk of an event may eventually reach 100% (as with all-cause mortality over time), it is generally more interesting to consider how quickly the event occurs. From a pharmacoepidemiologic perspective, there is interest in whether the protective or harmful effects of a drug occur immediately or only after some time, and whether exposure needs to be consistent or whether intermittent use of a drug will provide the same result. Because risks and rates are different, they cannot be compared directly. Despite this, the risk ratio and rate ratio are equivalent under some conditions. In particular, this assumption is reasonable when we consider time intervals short enough that the risk is 20% or less.

DATA SOURCES

Various data sources are used worldwide to conduct pharmacoepidemiologic research. Examples of these data sets are shown in Table 8. In general, the advantages of these data sets include size, validity of information, and ability to identify individuals who use resources. The weaknesses of these data sets tend to include limited generalizability (e.g., Medicaid); missing data, including use of over-the-counter or other uncovered medications and services, or when individuals seek care outside the network;

Table 8. Selected Examples of Databases Used in Pharmacoepidemiologic Research

U.S. Databases	Population Included
Group Health Cooperative	HMO members
HMO Research Network	Beneficiaries of member organizations
Kaiser Permanente	HMO members
Medicaid	Low-income individuals and families who meet Medicaid eligibility criteria
Medical Expenditure Panel Survey	Noninstitutionalized civilians in the United States
Medicare Current Beneficiary Survey	Older adults, and disabled or institutionalized individuals in the United States
National Health and Nutrition Examination Surveys (NHANES)	Adults and children in the United States; includes interviews and physical examination data
Other Databases	
Saskatchewan Health	Beneficiaries of the provincial health benefit
UK General Practice Research data set	Individuals who receive care in general practice settings in the United Kingdom

HMO = health maintenance organization.

and potential difficulty working with the data. In addition, obtaining permission to use these data sets may be difficult or expensive.

Data may also be reported spontaneously, as with the U.S. Food and Drug Administration's MedWatch, the Vaccine Adverse Event Reporting System, and the World Health Organization Programme for International Drug Monitoring, which is used for global monitoring. The value of these spontaneously reported event data sets lies in their size, potential for identifying these events (also known as signal detection), utility in hypothesis generation, and potential to contribute to clinical care. However, the link between any adverse event and a drug product may be subjective; hence, capture of these events is incomplete, making an estimation of incidence rates essentially impossible, and data may be missing even when an event is reported.

Registries

According to the Agency for Healthcare Research and Quality, a patient registry is "an organized system that uses observational study methods to collect uniform data (clinical and other) to evaluate specified outcomes for a population defined by a particular disease, condition, or exposure, and that serves one or more predetermined scientific, clinical, or policy purposes." These tools include detailed information about specific events in defined groups of individuals covering some specified amount of time. They can be used as simple ways to gather information and explore data, as links to populations, and to provide more detailed information about exposed individuals. One type of registry is that used to track the experience of individuals who received clozapine, which has been used to prevent reexposure among individuals who experience drug-related bone marrow suppression or leucopenia. The United States National Registry of Drug-Induced Ocular Side Effects is another example of this type of registry.

Another type of registry is exemplified by the Surveillance, Epidemiology, and End Results program, also known as SEER. This program is administered by the National Cancer Institute of the National Institutes of Health. It includes information on cancer epidemiology and survival from population-based cancer registries that cover more than 25% of the U.S. population. From a pharmacoepidemiologic perspective, this resource also includes data on the first course of treatment. In addition, one of the goals of this program is to assess the epidemiology of iatrogenic malignancies. Registries may also be linked between data sets to provide more information about specific groups of people. For example, the SEER-Medicare database was designed to focus on the experiences of older adults who have cancer.

Panel Statements and Guidelines

Many pharmacoepidemiologic studies make use of large databases for retrospective analyses. As a result, it makes sense to examine the ideal properties of these tools. These properties include whether the database is relevant to the question, how valid and reliable the data are, how data were linked between files, and how member eligibility was determined. Furthermore, studies using these resources should follow the precepts of good research design. For example, it is useful to know whether investigators set out to describe data to test specific hypotheses, whether the goal was to conduct careful analyses to help generate hypotheses, or whether a poorly conceived fishing expedition to find statistically significant associations was undertaken. In addition, the study design should reflect care and thoughtfulness, limitations should be identified and addressed, and the experience of a comparison or control group should be assessed, among other issues.

The International Society for Pharmacoepidemiology published a guideline for good pharmacoepidemiology practice. The document suggests minimal practices and procedures to be considered for high-quality pharmacoepidemiologic research. Topics addressed in this document include protocol development, responsibilities, personnel, facilities, resource commitment, contractors, study conduct, communication, adverse event reporting, and archiving.

BIAS

Error in the design, execution, or analysis of the study and data collected may be random or systematic. Systematic error is commonly referred to as bias. Several major classes of bias occur in pharmacoepidemiologic studies, including confounding, selection bias, and information bias. Other types of bias, such as immortal time bias, have been described as well. A primary difference between these types of error is that random errors will diminish as the size of the study increases, whereas systematic error does not diminish in that case. For example, the mean age of study participants may not reflect the mean age of individuals in the general population with the same condition. But, as the sample size approaches the number of people in the population of interest, the mean age of individuals in the sample will converge toward that value in the population.

Confounding

Confounding refers to circumstances in which the effects of many factors on an outcome are not separated. As a result, it is difficult to separate the individual contribution of these factors and, ultimately, measures of the effect of exposure on the outcome are distorted (Figure 12). For confounding to occur, the variable must be related to the exposure and the outcome. If the factor is on the causal pathway, however, it is not a confounder.

For example, imagine that there is interest in the effect of a drug on the occurrence of lung cancer but that the individuals who are or were smokers received the drug more often than individuals who never smoked. The result of uncontrolled confounding is that we are unable to tell what effect is caused by the factor of interest (drug use in this case) and what is caused by the confounder. One version of confounding of particular concern in pharmacoepidemiologic research is confounding by indication. This variant occurs when individuals who take a drug differ from individuals who do not, by virtue of the indication for that drug. For instance, individuals who have hyperlipidemia and take lipid-lowering drugs have higher average cholesterol concentrations than individuals who do not take this type of drug. The indication for lipid-lowering drugs makes the comparison of lipid-lowering drug exposure to non-exposure partly a comparison of high cholesterol and low cholesterol concentrations. If cholesterol concentration is prognostically related to whatever the study outcome was, then confounding by indication results. One approach to this issue is to focus on individuals who have the same underlying condition but who receive different treatments. This approach can also present problems and will likely not control all of this type of confounding because there may be fundamental issues that contribute to the choice of drugs, such as disease severity.

Another type of potential confounding that is important in pharmacoepidemiology occurs when medications or other health care interventions are used differentially among individuals with the same condition. This type of error is known as channeling bias or confounding by severity because the drug or other intervention is channeled to one group of people or away from another group of individuals, generally on the basis of the severity of the condition of interest or of comorbid illnesses. In addition, when severity of illness is an indication for a drug, this type of bias is considered confounding by indication. Channeling bias is thought to be common and under-recognized in pharmacoepidemiology. This is partly because this and other related biases, such as confounding by indication, are related to when, why, and for whom drugs are used. In many cases, this is a relatively subjective process that relies on the prescriber.

An example of channeling bias occurs when a new drug is used preferentially in individuals with more severe illness. If individuals at the end of life are less likely to be prescribed certain drugs, or are less likely to take

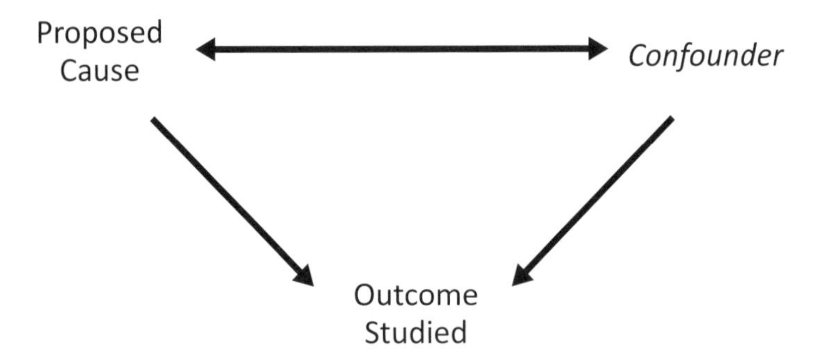

Figure 12. Relationship of a confounder to cause and outcome.

those drugs even if they are prescribed, exposure to the drugs in question will be imbalanced.

One such example of potential channeling bias in the biomedical literature involves use of the selective cyclooxygenase-2 nonsteroidal anti-inflammatory drugs. These drugs were originally marketed as being less likely to cause serious gastrointestinal ulceration and bleeding compared with nonselective nonsteroidal anti-inflammatory drugs. It was also believed that the selective drugs were more suitable for individuals at higher risk of these events. But, although the individuals who were prescribed the selective drugs may have been at increased risk of gastrointestinal toxicity, they were often also at a higher risk of cardiovascular events. As a result, channeling of the selective nonsteroidal anti-inflammatory drugs to older individuals to decrease one set of serious adverse effects may have contributed to another set.

Control of Confounding

Confounding is prevented or controlled in several ways, including randomization, restriction, stratification, and matching. Other tools used to prevent or control confounding are propensity scores and include the potential confounder as a covariate in a regression model. Randomization is a useful tool to prevent confounding because it removes the imbalance between groups, but it cannot always be used to study epidemiologic questions. Investigators can also restrict the study to include only individuals who have the same (or almost the same) value for a potential confounder. A disadvantage to restriction is that generalization of the study results may be limited. Stratifying the analysis to compare individuals in different groups with the same level of the potential confounder is another technique to help adjust for this bias. Matching involves finding individuals or groups with the same value for a particular variable and, when used in cohort studies, has the advantage of being intuitive to many clinicians because of its similarity to methods used in clinical trials. In case-control studies, matching is more difficult because it is based on the outcome.

Propensity scores are the conditional probability of exposure, given the combination of covariates observed, and can be used in modeling, matching, restriction, and stratification.

Although it is not appropriate to assess the potential for confounding by performing significance testing of factors related to exposure and outcome, the distribution of clinical and demographic characteristics can be examined in a well-designed table. In addition, formal assessment of confounding is done by comparing the crude estimate with the estimate after adjusting for the potential confounder. In general, if the crude estimate changes by at least 10%, the covariate is assumed to be a confounder.

Even when efforts to control confounding are made, residual confounding may still be present. This type of bias results for several reasons. First, when an analysis is stratified using broad strata of a potential confounder, confounding may occur within those levels of the covariate. Second, residual confounding may occur when covariates are not controlled or when there is error in covariate measurement. Residual confounding is generally controlled or prevented using narrow strata, avoiding the use of open-ended categories, choosing covariates to control for carefully, and addressing measurement errors.

Selection Bias

Selection bias occurs because of methods used to choose study participants and because of issues related to study participation. For example, if there are systematic differences in exposure and outcome between people who take part in a study and people who do not, this type of bias results. The underlying idea is that the association between exposure and outcome differs between study subjects and individuals who are eligible but do not take part.

In a pharmacoepidemiologic context, one way that selection bias occurs is when individuals in a study who are exposed to a particular drug drop out of the study twice as often as individuals who do not receive the drug. Another source of selection bias has to do with seeking medical attention. Individuals with a family history of some disease may be more likely to have regular checkups and be screened. Similarly, some symptoms are more bothersome than others, and if individuals with such symptoms seek treatment differentially, bias can result.

Information Bias

When data collected on study participants are wrong, the result may be misclassification of the exposure measure, the outcome measure, or both. An outcome measure that is equally misclassified among exposed and unexposed individuals is termed *non-differential misclassification.* Similarly, an exposure measure that is equally misclassified among people who have the study outcome and those who do not is non-differential misclassification.

Recall bias is a type of information bias that occurs when study participants imperfectly remember information about their exposure, particularly after the outcome has occurred. Although recall bias usually infers lost information, there are examples of improved reporting. For example, parents of children who became ill may have better recall than parents whose children did not experience the event. A difference exists between recall that is fuzzy and recall that is fuzzy differentially between exposed and unexposed groups. Biased follow-up can also produce differential misclassification if the outcome is recognized less often in unexposed individuals than in exposed individuals.

The effect of non-differential misclassification of a binary exposure measure (exposed/non-exposed) will result in bias toward a null effect, but non-differential misclassification of a categorical (high/medium/low) or continuous exposure measure is not predictable (could be toward or away from the H_0). Non-differential misclassification of a binary outcome measure (acute MI/no acute MI) does not result in bias either toward or away from the H_0.

Information bias can be avoided in several ways. Information can be collected in ways that improve recall, medical records can be used instead of patient interviews, alternative control groups can be used, and, in the case of biased follow-up, physical examinations can be included in the study design to bolster patient-reported medical histories. Because this bias is a result of misclassification, researchers can help minimize its occurrence and consequences by validating their data relative to a reference standard, such as a patient's medical chart.

Immortal Time Bias

This type of information bias is a result of including person-time in an analysis when no outcome could have occurred. For example, in the outcome of hospitalization, study participants do not contribute person-time while in the hospital. Because eligibility for the outcome is a necessary condition for measures such as incidence density to be valid, including observation time from individuals who are not at risk of the event—and thus, who cannot have the event—in the calculation artificially inflates the denominator. As a result, the intervention might look safer than it really is.

Effect Modification

This phenomenon is not necessarily a bias, but it is often discussed in the same context. Effect modification occurs when the effect of a covariate on an outcome depends on the level of at least one other variable. For example, if the effect of age on postoperative pain depends on the individual's sex, effect modification is present. Also of interest may be understanding who benefits from most public health campaigns, such as vaccination against common infectious diseases.

COMORBIDITY INDICES

These measures of burden caused by concurrent illness quantify the effect of many conditions on outcome. The premise underlying comorbidity indices is simple: individuals with more than one condition generally have worse clinical, economic, and patient-reported outcomes than those who are in otherwise good health. This idea is analogous to comparing two patients, one who is an aerobically trained athlete and one who has hypertension, has diabetes mellitus, and is morbidly obese. If each of these individuals undergoes the same surgical procedure, the otherwise healthy person is more likely to recover faster, regain more baseline function with time, have a shorter stay in the hospital, and so on.

Many comorbidity indices are used in research and practice. Among the most commonly used of these tools are the Acute Physiology and Chronic

Health Evaluation II (APACHE II) and the Charlson Comorbidity index. The APACHE II is typically used to provide a measure of overall disease severity among patients in intensive care units, although it is also used in other patient populations. Its score is a function of commonly monitored physiologic values assessed during the person's first 24 hours in the intensive care unit, plus age and prior health status. The APACHE II score can then be used to stratify patients and help clinicians and researchers, as well as contribute to understanding how medical resources are used.

Two recent publications illustrate the use of the APACHE II. In the first, the APACHE II was used to assess the risk of delirium in a study of lorazepam use among 198 adult, mechanically ventilated individuals without neurologic disease who were admitted to medical or coronary intensive care units. The investigators found that lorazepam use and APACHE II scores were independent risk factors for developing delirium and, specifically, that each unit increase in the APACHE II score was associated with a 6% increase in the probability of developing this condition.

Investigators have also used the APACHE II index to estimate survival in hospitalized individuals outside the intensive care unit. Specifically, researchers hypothesized that this instrument is useful to predict long-term mortality in individuals with chronic obstructive pulmonary disease admitted to medical units. In this study, logistic regression and Cox proportional hazards models were used to estimate the relationship between APACHE II score, survival, and length of hospitalization, adjusting for clinical and demographic characteristics. In this study, increasing the APACHE II score was associated with a 75% rise in long-term mortality when evaluated using proportional hazards methods and with increased mortality at 3 years using a logistic model (OR = 2.62; 95% CI, 1.12–6.16). Length of stay was not statistically significantly associated with APACHE II score.

The Charlson index uses weighted comorbidity data to estimate 1-year mortality. With definitions from the *International Classification of Diseases,* 9th Revision, *Clinical Modification* (ICD-9-CM), researchers have adapted this tool; scores have explained the results of different medical or surgical procedures. Investigators have also found that different adaptations of the Charlson index typically agree in the choice of ICD-9-CM code for a given comorbid condition, that these iterations generally produce concordant estimates of the overall score, and that this tool can be used to predict groups of individuals who are at increased risk of dying in the hospital and after hospitalization.

Other techniques have also been used to help adjust for baseline risk, including pharmacy models. The Rx Risk model is one such effort. In this technique, investigators developed a model in which prescription drugs were separated into classes of illnesses to identify individuals with chronic conditions and to explain or predict health care costs. This type of approach was useful in explaining the central 60% of the cost distribution. Pharmacoepidemiologic applications for this model may include other administrative data sets, particularly when diagnostic codes are unavailable.

STATISTICS AND PHARMACOEPIDEMIOLOGY

Earlier, biostatistics was discussed in depth. To avoid unnecessary duplication of material, this discussion focuses on some issues particularly applicable to epidemiologic research.

Often, the term *statistical significance* is treated as some absolute indicator of truth, but it tells us only that the p value is less than the chosen α level. Because that parameter is chosen either arbitrarily or, in biomedical research, by convention, the meaning of statistical significance also is variable.

Researchers sometimes perform statistical significance testing to assess whether participants' characteristics at baseline are balanced. The idea behind this practice is that if observed differences do not reach statistical significance, no confounding occurs because of that variable, but this practice and interpretation does not tell us what we want to know for several reasons.

First, differences in the distribution of patient characteristics do not tell us anything about the amount of confounding. Of the needed conditions for a factor to be considered a confounder, the observed imbalance between groups on that variable is just one. Second, an assessment of confounding requires an understanding of the degree of association between the factor, exposure, and outcome, but statistical significance testing depends on the sample size and degree of association being evaluated. Third, in a randomized study, it is critically important to understand what H_0 is being tested when this practice is used. Because the H_0 is that any observed differences are attributable to chance, if the H_0 is rejected, the interpretation must be that randomization did not work. As a result, it makes much more sense to examine the relationship between the factor, exposure, and event. Methods to accomplish this include descriptive techniques, such as graphs and tables, as well as correlation and regression models.

DIAGNOSTIC TEST PERFORMANCE

A variety of summary measures are useful in understanding the contribution of epidemiology to the understanding of pharmacotherapy. Diagnostic methods, sensitivity, specificity, and predictive value are all related to identifying individuals with the outcome of interest. The first step is to clearly define what constitutes the outcome compared with something that looks similar but is not the event.

When a clear definition of the event has been developed, it is useful to estimate the sensitivity, or percentage of individuals who had the outcome and were correctly identified, and the specificity, or proportion of individuals who did not have the outcome and were correctly identified. Ideally, a test or analysis that is 100% sensitive and 100% specific is preferred because all events identified will be correct, no events will be missed, and no events will be incorrectly identified. Most methods do not accomplish this goal. As a result, clinicians and researchers must decide on the optimal tradeoff. For example, although a correct reading is always preferred, in some cases, we may be willing to accept a higher chance of a false-positive result, whereas at other times, a higher probability of a false-negative reading may be acceptable. Once this task is accomplished, a plot of sensitivity and 1-specificity can be constructed to give the receiver operating characteristic curve. The area under this curve provides an estimate of how the research methods perform compared with chance.

We may also be interested in assessing how well a test or method is able to identify individuals with or without the condition of interest. These measures are called the predictive value positive and predictive value negative. These estimates are calculated similarly to sensitivity and specificity, but instead of comparing the gold standard estimate with the test performance, we start with the new method and compare those results with the objective measure. For example, imagine that 100 individuals had an adverse drug reaction and that 75 of them were correctly identified. The sensitivity of the method used is 75%. Similarly, if 100 individuals did not have a particular event and 75 are identified correctly, the specificity is 75%.

In contrast, by comparing results with the gold standard, the predictive value measures can be predicted. Imagine that we would like to know how well a test predicts the presence of early-stage cancer. If 50 people have a positive test and 25 are later found to have that malignancy, the predictive value positive will be 50%. If 100 individuals had a negative test result and 90 of them did not have the condition, the predictive value negative would be 90%.

DATA PRESENTATION

Throughout the process of planning and carrying out research, data presentation must be considered. This issue is also important for the individuals reading the results because interpreting the validity of pharmacoepidemiology research relies heavily on how well the investigator has addressed bias—particularly confounding, information bias, and selection bias.

The Consolidated Standards of Reporting Studies statement, updated in 2001, offers additional tools to help researchers in presenting data and to assist readers in assessing how studies were carried out and whether the estimates are valid. This statement was designed to improve the quality of parallel-group randomized study reports, but its suggestions are also applicable to pharmacoepidemiologic research.

The Consolidated Standards of Reporting Studies statement includes two parts: a checklist for authors to use in presenting research results and a flowchart for readers. The 22-item checklist is subdivided into the main sections of a paper. The items in each section pertain to issues that should be addressed in that part of the paper. The flowchart indicates how many individuals were assessed for inclusion in the study, excluded from participating, randomized and assigned to each treatment group, lost to follow-up, and included in the analysis.

Another, more general approach to clear, useful data presentation emphasizes ways to produce simple, informative tables and graphs without misleading the reader or cluttering the page. The importance of making liberal use of methods such as these to improve the communication of research results cannot be overstated, particularly as the number of journals and rate of publication increases.

SUMMARY

In summary, interpreting the medical literature requires a basic understanding of biostatistics together with the types of statistical tests used to summarize or make inferences from data. Once mastered, these skills can be applied to both randomized, controlled clinical trials and pharmacoepidemiologic studies. If pharmacists are to contribute to the improvement of health outcomes, they must be able to apply the medical literature to the patient populations and individual patients they serve.

SELECTED BIBLIOGRAPHY

Abenhaim L, Moride Y, Brenot F, et al. Appetite-suppressant drugs and the risk of primary pulmonary hypertension. International Primary Pulmonary Hypertension Study Group. N Engl J Med 1996;335:609–16.

Begg C, Cho M, Eastwood S, et al. Improving the quality of reporting of randomized controlled trials: the CONSORT statement. JAMA 1996;276:637–9.

Bradford-Hill AB. The environment and disease: association or causation? Proc R Soc Med 1965;58:295–300.

Craig BM, Kreling DH, Mott DA. Do seniors get the medicines prescribed for them? Evidence from the 1996–1999 Medicare Current Beneficiary Survey. Health Aff 2003;22:175–82.

Dillon CF, Hirsch R, Rasch EK, Gu Q. Symptomatic hand osteoarthritis in the United States. Prevalence and functional impairment estimates from the third US National Health and Nutrition Examination Survey, 1991–1994. Am J Phys Med Rehabil 2007;86:12–21.

Gehlbach SH. Interpreting the Medical Literature, 4th ed. New York: McGraw-Hill, 2002.

Glynn RJ, Knight EL, Levin R, Avorn J. Paradoxical relations of drug treatment with mortality in older persons. Epidemiology 2001;12:682–9.

Johnson ES, Koepsell TD, Reiber G, Stergachis A, Platt R. Increasing incidence of serious hypoglycemia in insulin users. J Clin Epidemiol 2002;55:253–9.

Moher D, Schulz KF, Altman D; for the CONSORT group. The CONSORT statement: revised recommendations for improving the quality of reports of parallel-group randomized studies. JAMA 2001;285:1987–91.

Motheral B, Brooks J, Clark MA, et al. A checklist for retrospective database studies—report of the ISPOR Task Force on Retrospective Databases. Value Health 2003;6:90–7.

Peterson AM, Nau DP, Cramer BS, Benner J, Gwadry-Sridhar F, Nichol M. A checklist for medication compliance and persistence studies using retrospective databases. Value in Health, published early online 11/22/2006. DOI: 10.1111/j.1524-4733.2006.00139.x.

Petri H, Urquhart J. Channeling bias in the interpretation of drug effects. Stat Med 1991;10:577–81.

Rothman KJ. Epidemiology: An Introduction. New York: Oxford University Press, 2002.

Rothman KJ, Greenland S, eds. Modern Epidemiology, 2nd ed. Philadelphia: Lippincott-Raven Publishers, 1998.

Schneeweiss S, Glynn RJ, Tsai EH, Avorn J, Solomon DH. Adjusting for unmeasured confounders in pharmacoepidemiologic claims data using external information: the example of COX2 inhibitors and myocardial infarction. Epidemiology 2005;16:17–24.

Strom BL, ed. Pharmacoepidemiology, 4th ed. Hoboken, N.J.: John Wiley & Sons, 2005.

Suissa S. Effectiveness of inhaled corticosteroids in chronic obstructive pulmonary disease. Immortal time bias in observational studies. Am J Respir Crit Care Med 2003;168:49–53.

Tufte ER. Envisioning Information. Cheshire, Conn.: Graphics Press, 1990.

Tufte ER. Visual Explanations: Images and Quantities, Evidence and Narrative. Cheshire, Conn.: Graphics Press, 1997.

Tufte ER. The Visual Display of Quantitative Information, 2nd ed. Cheshire, Conn.: Graphics Press, 2001.

Tufte ER. Beautiful Evidence. Cheshire, Conn.: Graphics Press, 2006.

Zuvekas SH, Cohen JW. Prescription drugs and the changing concentration of health care expenditures. Health Aff 2007;26:249–57.

SELF-ASSESSMENT QUESTIONS

1. In a study of the number of adverse drug events in a group of people over a year, the authors report that five events occurred per 10,000 person-years. Which one of the following epidemiologic measures does this estimate represent?

   A. Risk
   B. Hazard ratio
   C. Prevalence
   D. Incidence rate

2. A study examined emergency department admissions in a city secondary to recreational drug use. Assume the researchers are able to capture data on all such admissions in a given time. Which one of the following epidemiologic measures do these data represent?

   A. Period prevalence
   B. Prevalence
   C. Incidence
   D. Incidence density

3. You are the pharmacist on the Pharmacy and Therapeutics Committee for a large health maintenance organization. At the meeting last month, you were asked to review the pharmacoepidemiologic evidence regarding the incidence of an adverse effect. Episodes of this adverse effect require affected individuals to miss several days of work. Which one of the following measures will be most useful to the companies whose employees receive health care from your organization?

   A. Relative risk reduction
   B. Absolute risk reduction
   C. Incidence
   D. Prevalence

4. Investigators of a study observe the experience of a group of people for a defined amount of time. The observation period is started after the event of interest has already occurred, and the study participants are observed to determine exposure status. The incidence rate in individuals who were exposed to the risk factor of interest is then

compared with the incidence rate in individuals not exposed to that risk factor. Which one of the following is this type of study?

A. Nested case-control
B. Prospective cohort
C. Retrospective cohort study
D. Prospective case-control

5. You are the pharmacist on the hospital Pharmacy and Therapeutics Committee. The committee is considering whether to take a drug off the formulary. The senior members of the committee are concerned because recent case reports in the biomedical literature have suggested that the use of this drug increases the risk of a rare but serious musculoskeletal event. Which one of the following study designs would you consider most suitable for testing to see whether this concern is well founded?

   A. Randomized, controlled study
   B. Case-control study
   C. Cohort study
   D. Case-cohort study

6. Which one of the following characteristics is the main reason that many health services researchers prefer to use administrative and clinical databases to conduct studies?

   A. Ability to examine large populations for extended observation periods
   B. Ability to study unbiased samples
   C. Ability to generate results superior to those obtained from randomized, controlled studies
   D. Ability to easily impute valid estimates for variables when data are missing

Questions 7 and 8 pertain to the following case.

Using a large database, a cohort study of individuals without health care insurance is conducted to assess the rate of adverse drug events resulting in hospitalization using a large database. All hospitals in this multicounty area have agreed to contribute their data to this data set.

7. Which one of the following is a main limitation of this data set?
   A. Confounding will interfere with interpretation of the study results.
   B. It may not be possible to generalize the study findings to individuals in the same area who have insurance.
   C. Selection bias will interfere with interpretation of the findings.
   D. Information bias will interfere with interpretation of the findings.

8. In conducting the analyses, you find that the data for several of the variables of interest are misclassified but that there is no apparent pattern to the errors. Which one of the following can be done to address this limitation?
   A. The study population and analyses can be stratified.
   B. The study population and analyses can be limited to validated groups.
   C. The analyses can be restricted to include only individuals who received all of their care from just one participating facility.
   D. The analyses can be restricted to include only individuals who received all of their care from participating facilities.

9. Researchers sought to estimate the occurrence of birth defects related to in utero exposure from drug A. They interviewed women who had recently given birth at a large tertiary care center, reviewed prescription drug dispensing records for the study participants, and examined medical charts of the infants to see whether there was any indication of a birth defect. Which one of the following statements is the best description of potential bias in this retrospective cohort study?
   A. Estimates are unbiased because population-based cohort studies provide valid and reliable data.
   B. Information bias is the most important potential bias because the number of cases and total at-risk population may be incompletely observed.
   C. Estimates from this study are unbiased because birth defects are generally uncommon, and the researchers had access to information from several sources.
   D. Confounding is the most significant potential bias of these estimates.

10. As part of an institution-specific effort to improve the quality of health care delivered, a colleague wants to study the incidence of adverse drug

effects using clinical and administrative data from all individuals admitted to the hospital in the previous year. As the pharmacy liaison to the investigational review board, you have been asked to review the study protocol. You are concerned that this data source may be biased. Which one of the following types of bias is most important in considering this research topic?

A. Selection bias
B. Information bias
C. Confounding
D. Ecological bias

11. An observational study of the effects of a class of drugs indicates that individuals in one group who use these drugs are more likely to develop kidney dysfunction than are people who do not use these drugs. If these drugs are used to treat a variety of underlying conditions and each condition is related to the development of kidney dysfunction, which one of the following types of bias may exist in this study?

    A. Immortal time bias
    B. Diagnostic bias
    C. Confounding by indication
    D. Information bias

12. In a randomized, controlled study, the authors include a flowchart of study participants. This figure indicates that twice as many individuals randomized to receive drug A dropped out of the study compared with study participants randomized to receive drug B. Which one of the following potential biases does this observation represent?

    A. Confounding
    B. Information bias
    C. Immortal time bias
    D. Selection bias

Questions 13 and 14 pertain to the following case.
In conducting a cohort study, you find evidence that the data set and independent drug dispensing data disagree for 10% of individuals.

13. Assuming that the drug dispensing data are correct, which one of the following pieces of information is most important as you decide what to do next?

    A. Does not matter; such a high degree of misclassification that none of the study results will be believable
    B. How the misclassified individuals were distributed between exposed and unexposed groups
    C. How the misclassified individuals were distributed by sex
    D. How the misclassified individuals were distributed by age

14. On further exploration, you find that of the individuals misclassified, 50% received the drug of interest but were classified as being unexposed, and 50% were classified as being unexposed but actually received the drug. Outcomes were similar for exposed and unexposed individuals, however. Which one of the following types of misclassification is this and why?

    A. Non-differential because the misclassification is not related to the outcome
    B. Differential because the misclassification is only related to the exposure
    C. Non-differential because the data are similarly prone to error for exposed and unexposed individuals
    D. Differential because exposed and unexposed individuals are misclassified

15. You are analyzing data from a case-control study of the effect of drug A on cardiovascular events. As you review the data, you realize that some individuals were classified as having received the drug when they did not and that the distribution of these individuals is similar between cases and controls. Which one of the following is the likely consequence of this misclassification?

    A. Estimate biased either away from or toward the null
    B. No effect
    C. Estimate biased away from the null hypothesis of no effect
    D. Estimate biased toward the null hypothesis of no effect

16. In a study of a lipid-lowering drug in 100 individuals, 15 individuals who received the drug were identified as cases when they did not meet the study criteria for that event, and 15 other individuals who did not receive the drug had events that should have qualified them as cases. Which one of these approaches would be possible methods for correcting this situation?

    A. Matching
    B. Restriction
    C. Not a problem, do nothing
    D. Linking data sources together

17. Imagine that the outcome of a study is rare but potentially serious. Because it is so rare, imagine that it is difficult to recognize. If the misclassified individuals were all in the outcome group, not the exposure group, which one of the following could be used to address this situation?

    A. There is no need to do anything because it is unknown whether the misclassification was differential.
    B. Additional testing is needed to rule out other possible causes of this outcome.
    C. Randomization is needed.
    D. Additional interviews are needed to collect more information from patients.

Questions 18 and 19 pertain to the following case.

You and your colleagues conduct a randomized, double-blind, controlled study of a drug. In constructing the preplanned tables, you find imbalances in the distribution of clinical and demographic characteristics between the experimental groups.

18. Which one of the following types of bias are you concerned about?

    A. Confounding
    B. Immortal time bias
    C. Selection bias
    D. Information bias

19. Which one of the following argues against the presence of channeling bias?

A. The imbalanced factors are in the causal pathway.
B. The study is randomized.
C. The imbalanced factors are related to the exposure.
D. The imbalanced factors are related to the outcome.

20. The research team you are part of conducts a large cohort study to estimate the risk of cardiovascular events associated with use of a new drug. The rate ratio estimated from this study suggests that using this drug was associated with a 20% increase in the rate of events compared with not using the drug. A member of the project team notes that this estimate is substantially lower than that found by other researchers. Which one of the following explanations best explains the difference between your findings and the other published estimates?

    A. Differential misclassification
    B. Channeling bias
    C. Confounding
    D. Non-differential misclassification

21. You are studying the relationship between the use of decongestants and hemorrhagic stroke, a rare event in young, otherwise healthy women. Which one of the following study designs is most appropriate?

    A. Cohort
    B. Case-control
    C. Randomized, controlled trial
    D. Prospective

22. The first table in many articles published in the biomedical literature generally shows the clinical and demographic characteristics of study participants. Which one of the following potential biases does this type of information address?

    A. Confounding
    B. Selection bias
    C. Information bias
    D. Effect modification

23. A Consolidated Standards of Reporting Studies–style flowchart is recommended to be included in studies to address which one of the following types of bias?

A. Immortal time bias
B. Selection bias
C. Information bias
D. Confounding

Questions 24 and 25 pertain to the following case.

You are the pharmacoepidemiology specialist in your health care organization. You conduct a study of the occurrence of respiratory disease relative to the use of a drug among the patients in your clinic and obtain the following results:

- Incidence of respiratory disease among exposed patients = 200/100,000 person-years
- Incidence of respiratory disease among unexposed patients = 20/100,000 person-years
- Incidence in the general population = 21/100,000 person-years

24. Which one of the following is the interpretation of the rate ratio for this population?
    A. Exposed individuals have a risk of developing respiratory illness that is one-tenth as high as their unexposed colleagues.
    B. Exposed individuals have a risk of developing respiratory illness that is 10 times higher than their unexposed colleagues.
    C. Exposed individuals have a risk of developing respiratory illness that is 5% higher than their unexposed colleagues.
    D. Exposed individuals have a risk of developing respiratory illness that is 5% lower than their unexposed colleagues.

25. Which one of the following is the interpretation of the attributable fraction for these data?
    A. 90% of the risk of respiratory illness in exposed individuals is caused by the exposure.
    B. 20% of the risk of respiratory illness in exposed individuals is caused by the exposure.
    C. 5% of the risk of respiratory illness in exposed individuals is caused by the exposure.
    D. 9% of the risk of respiratory illness in exposed individuals is caused by the exposure.

ANSWERS TO SELF-ASSESSMENT QUESTIONS

BASICS OF BIOSTATISTICS

1. Answer: C

Answer C (median) is correct because the median is an estimate of the center of the distribution of responses and is not sensitive to extremes. In particular, the median is a measure of average that can be used for ordinal data. Answer A (mean) is incorrect, because when the responses are ordered, it cannot be assumed that the distance between values is uniform. In addition, the mean is subject to distortion by extreme values, and satisfaction is a crude indicator of the quality of health care. Answer B (interquartile range) is incorrect because this measure reflects the distribution of the central 50% of the data. Answer D (mode) is incorrect because the mode is the most common value.

1. Rosner B. Fundamentals of Biostatistics, 6th ed. New York: Duxbury Press, 2005.
2. Miaskowski C, Nichols R, Brody R, Synold T. Assessment of patient satisfaction utilizing the American Pain Society's Quality Assurance Standards on acute and cancer-related pain. J Pain Symptom Manage 1994;9:5–11.

2. Answer: B

Answer B is correct because CIs indicate the range over which a point estimate is consistent with the data and indicate the random error in an estimate. The p value is the probability, conditional on the H_0, of observing at least as large a difference as the difference observed. The p values are calculated using statistical models that correspond to the type of data collected. In this question, a difference from 1 unit to 3.5 units on the 0–10 numeric rating scale is consistent with the data. Answer A is incorrect because the p value does not provide an estimate of the magnitude of the observed difference, although it is often interpreted this way. Answer C and Answer D are also incorrect because neither a precise p value nor the CI provides insight sufficient for clinical decision-making.

1. Rosner B. Fundamentals of Biostatistics, 6th ed. New York: Duxbury Press, 2005.
2. Rothman KJ. Epidemiology: An Introduction. New York: Oxford University Press, 2002.

3. Answer: D

Answer D is correct because one of the treatments was statistically better than the other, but we cannot tell which one from the information given. Answer A is incorrect because we cannot tell whether the first treatment provided better management of postoperative pain than the second treatment. Answer B is incorrect because we cannot tell whether the second treatment provided better management of postoperative pain than the first treatment. Answer C (no difference) is incorrect because a statistically significant difference was observed between the two treatments.

1. Rosner B. Fundamentals of Biostatistics, 6th ed. New York: Duxbury Press, 2005.
2. Rothman KJ. Epidemiology: An Introduction. New York: Oxford University Press, 2002.

4. Answer: A

Answer A is correct because a p value less than or equal to the prespecified α is interpreted as statistically significant. Answer B (p value is an absolute indicator) is incorrect because the p value does not indicate which treatment was better. Answer C (statistical significance is independent) is incorrect because the assessment of statistical significance depends on the chosen significance level. Answer D (relative importance of statistical significance) is incorrect because statistical significance is not necessarily more important than clinical significance.

1. Rothman KJ. Epidemiology: An Introduction. New York: Oxford University Press, 2002.
2. Gottschalk A, Smith DS, Jobes DR, Kennedy SK, Lally SE, Noble VE, et al. Preemptive epidural analgesia and recovery from radical prostatectomy: a randomized controlled trial. JAMA 1998;279:1076–82.

5. Answer: C

Answer C is correct because larger differences between treatments are generally needed to reach clinical significance than statistical significance. However, in statistically underpowered studies, it is possible for a difference to be clinically significant without reaching statistical significance. For example, in a clinical trial of antihypertensive pharmacotherapy, a difference between treatments of 1 mm Hg may be statistically significant; however, it is of questionable clinical significance for individual patients.

Answer A (clinical significance was less important than statistical significance) is incorrect because clinical significance is not necessarily less important than statistical significance, in this case or any other. Appropriate use of clinical significance and statistical significance relies on the interdependence between these data. Similarly, a large, clinically significant difference between groups may be observed, but if study design requirements are not met, the study results will unlikely affect practice. Answer B (differences for clinical significance are smaller than for statistical significance) is incorrect because the statement is untrue. The difference between treatments required for clinical significance is likely to be larger, not smaller, than that needed for statistical significance. Answer D (equality of clinical and statistical significance) is incorrect because clinical and statistical significance are not necessarily equally important.

1. Rothman KJ. Epidemiology: An Introduction. New York: Oxford University Press, 2002.
2. Gottschalk A, Smith DS, Jobes DR, Kennedy SK, Lally SE, Noble VE, et al. Preemptive epidural analgesia and recovery from radical prostatectomy: a randomized controlled trial. JAMA 1998;279:1076–82.

6. Answer: B

Answer B (an OR of 1.5 with a 95% CI that extends from 1.2 to 2.0) is the correct interpretation of this estimate because the data are consistent with ORs from 1.2 to 2.0, as shown in the CI. An OR of 1.2 is interpreted as a 20% increase in the likelihood that exposed individuals had the outcome, whereas an OR of 2.0 is interpreted as a 100% increase in the likelihood that exposed individuals had the outcome compared with unexposed individuals. Answer A (OR of 1.5 = 150% increase in likelihood of the outcome) is incorrect because an OR of 1.5 is not interpreted as a 150% increase in the likelihood of the outcome. An OR of 2.5 would represent a 150% increase, or an increase of 2.5 times the baseline odds. Answer C (an OR greater than 1.0 represents a decrease in the likelihood of the outcome) is incorrect because ORs above 1.0 are interpreted as increases in the likelihood of experiencing the outcome. Odds ratios below 1.0 are interpreted as a decrease in the likelihood of experiencing the outcome. Answer D (the estimated OR did not reflect a statistically significant result) is incorrect because the CI for the OR did not cross 1.0.

1. Kleinbaum DG, Klein M, Pryor ER. Logistic Regression: A Self-Learning Text, 2nd ed. New York: Springer, 2005.

2. Kleinbaum DG, Kupper LL, Muller KE, Nizam A. Applied Regression Analysis and Multivariable Methods, 3rd ed. New York: Duxbury Press, 1998.

7. Answer: A

	Outcome Yes	Outcome No	Total
Drug A	29	16	45
Drug B	14	31	45
Total	43	47	90

Answer A (OR = 4.0) is correct because the OR is estimated by cross-multiplying and dividing the observed number of events, which are the four data cells of the table, not using the row or column totals at all. In this case, this calculation is $ad/bc = (29 \times 31)/(16 \times 14) = 4.01$. Answer B, Answer C, and Answer D are simply incorrect calculations.

1. Rothman KJ. Epidemiology: An Introduction. New York: Oxford University Press, 2002.
2. Kleinbaum DG, Klein M, Pryor ER. Logistic Regression: A Self-Learning Text, 2nd ed. New York: Springer, 2005.

8. Answer: D

Answer D (the odds of having the outcome among individuals who received treatment A are 400% higher than in those in individuals who received treatment B) is correct because the OR is an estimate of the odds of the outcome in individuals in one treatment group compared with individuals who received the other treatment. This answer represents the correct interpretation of an OR of 4.0. Answer A (the odds of having the outcome among individuals who received treatment A are 10% higher than the odds in individuals who received treatment B) is incorrect because an OR of 1.1 is consistent with 10% higher odds of having the outcome among individuals who received treatment A than the odds of the outcome in individuals who received treatment B. Answer B (the odds of having the outcome are 75% lower among individuals who received treatment A than those among individuals who received treatment B) is incorrect because this answer is consistent with an OR of 0.25. Answer C (the odds of having the outcome among individuals who received treatment B are 10% lower than the odds among individuals who received treatment A) is incorrect because this is consistent with an OR of 1.1.

1. Rothman KJ. Epidemiology: An Introduction. New York: Oxford University Press, 2002.
2. Kleinbaum DG, Klein M, Pryor ER. Logistic Regression: A Self-Learning Text, 2nd ed. New York: Springer, 2005.

9. Answer: D

Answer D (treatment of hypertension with low doses of diuretics is from 3% to 23% more effective than treatment with β-blockers) is a correct interpretation of the CI, and it accounts for the plausible range of estimates that are consistent with the data and expressed in the CI. Answer A (treatment of hypertension with low-dose diuretics was 14% more effective in preventing cardiovascular disease events than treatment with β-blockers) is not as correct as Answer D because the former estimate does not account for the CI. Answer B (treatment of hypertension with β-blockers was 14% more effective in preventing cardiovascular disease events than treatment with low doses of diuretics) is incorrect because it represents an incorrect interpretation of the results. Specifically, this meta-analysis did not find that β-blockers were more effective than low-dose diuretics. Answer C (the difference observed between treatment with β-blockers and low doses of diuretics is not statistically significant) is incorrect; the difference observed between treatment with β-blockers and low doses of diuretics is, in fact, statistically significant, because the CI for the OR does not cross 1.0.

1. Psaty BM, Lumley T, Furberg CD, Schellenbaum G, Pahor M, Alderman MH, et al. Health outcomes associated with various antihypertensive therapies used as first-line agents: a network meta-analysis. JAMA 2003;289:2534–44.
2. Rothman KJ, Greenland S. Modern Epidemiology, 2nd ed. Philadelphia: Lippincott-Raven Publishers, 1998.

10. Answer: B

Answer B (5.4 mm Hg) is correct because it is the arithmetic mean of the differences between the blood pressure readings before and after study participants began using oral contraceptives. Answer A (6.5 mm Hg) is incorrect because it is the median of the differences observed. Answer C (7.0 mm Hg) is incorrect because it is the mode of the observed differences in blood pressure. Answer D (7.5 mm Hg) is incorrect because it is the 75th percentile value of the observed differences in blood pressure.

1. Rosner B. Fundamentals of Biostatistics, 6th ed. New York: Duxbury Press, 2005.
2. Altman DG, Bland JM. Quartiles, quintiles, centiles, and other quantiles. BMJ 1994;309:996.

11. Answer: D

Answer D (1.15) is correct because the SEM is calculated by dividing the SD of the posttreatment minus pretreatment SBP differences for each patient (4.6) by the square root of the number of study participants (n=16; = 4). Answer A (5.29) is incorrect because this value is obtained by dividing the square of the SD by the square root of the number of study participants. Answer B (1.32) is incorrect because this value is obtained by dividing the square of the SD by the number of study participants. Answer C (0.29) is incorrect because this value is obtained by dividing the SD by the number of study participants.

1. Rosner B. Fundamentals of Biostatistics, 6th ed. New York: Duxbury Press, 2005.
2. Altman DG, Bland JM. Standard deviations and standard errors. BMJ 2005;331:903.

12. Answer: C

Answer C (paired t-test) is correct because this test is used when each data point of the first sample is matched and related to a unique data point in the second sample. Paired samples may represent two sets of measurements on the same people, as when each person serves as his or her own control, or on different people who have been matched on criteria to be similar to each other. Answer A (chi-square) is incorrect because the chi-square test is used to evaluate observed differences when data are categorical. Answer B (t-test) is incorrect because the data are not independent. Answer D (Cox proportional hazards) is incorrect because Cox methods are used to assess time-to-event questions.

1. Rosner B. Fundamentals of Biostatistics, 6th ed. New York: Duxbury Press, 2005.
2. Kleinbaum DG, Klein M. Survival Analysis: A Self-Learning Text, 2nd ed. New York: Springer, 2005.

13. Answer: B

Answer B (6.1, 13.9) is correct because this interval was obtained by multiplying the SEM by 1.96, subtracting it from the observed mean difference to generate the lower limit of the 95% CI, and adding it to the observed mean difference to generate the upper limit of the 95% CI. Answer A (8.0, 12.0) is incorrect because this interval was derived by adding and subtracting 1.96 to and from the observed mean difference, respectively. Answer C (4.0, 8.0) is incorrect because this interval was constructed by adding and subtracting 1.96 to and from the observed median difference, respectively. Answer D (2.1, 9.9) is incorrect because this interval was constructed by multiplying 1.96 by the SEM and then adding and subtracting that quantity to and from the observed median difference, respectively.

1. Rosner B. Fundamentals of Biostatistics, 6th ed. New York: Duxbury Press, 2005.
2. Altman DG, Bland JM. Standard deviations and standard errors. BMJ 2005;331:903.

14. Answer: C

For describing data that are continuous, measures of central tendency, such as the mean (Answer C), give a good overall description of the data. Using the SEM (Answer A) to describe the age of a group of patients is not a good choice because this statistic measures the variability associated with different samples and does not represent the age of patients using the anticoagulation services. T-tests and chi-square tests (Answer B and Answer D) are used not as descriptive statistics but as inferential statistics, and they do not provide summary descriptions of a set of data.

1. Gaddis ML, Gaddis GM. Introduction to biostatistics: part 2, descriptive statistics. Ann Emerg Med 1990;19:309–15.
2. BMJ Publishing Group. Statistics at Square One. Available at *http://bmj.bmjjournals.com/collections/statsbk/index.shtml*. Accessed January 13, 2011.

15. Answer: A

The type of data in this question are “yes” or “no.” In this case, patients’ allergy information was either included (yes) or not included (no) on their admission orders. Data of this type are considered nominal (Answer A). Ordinal data (Answer B) are ranked in an order, although the differences between ranks are not consistent. Interval and ratio data (Answer C and

Answer D) are continuous data and are not used to classify dichotomous outcomes such as whether a patient's allergy status has been recorded.

1. Riegelman RK. Studying a Study and Testing a Test—How to Read the Medical Evidence, 4th ed. Philadelphia: Lippincott Williams & Wilkins, 2000.
2. BMJ Publishing Group. Statistics at Square One. Available at *http://bmj.bmjjournals.com/collections/statsbk/index.shtml*. Accessed January 13, 2011.

16. Answer: C

Cohort studies usually report the results of the comparisons between cohorts as RRs. This ratio represents the risk associated with an exposure compared with no exposure (or another exposure). Together with an estimate of the RRs, authors also should report an associated CI. If this CI includes the value of 1 within its bounds, then the possibility that the RR is different from 1 (i.e., that the risks are the same) cannot be excluded. Because the 95% CI in this question ranges from 0.65 to 1.07 and includes the value of 1, you can conclude that the RR estimate of 0.92 is not a statistically significant difference from equal risk (RR = 1, or the same risk in the numerator and denominator). A p value of less than 0.05 or less than 0.01 (Answer A and Answer B) are counted as statistically significant. Because the CI does not quantify the p value, it can only be said that it is not statistically significant. Although both a p value of greater than 0.05 or greater than 0.10 (Answer C and Answer D) represent nonsignificant results, readers are limited to concluding that the p value is greater than 0.05 (Answer C) on the basis of the CI.

1. Riegelman RK. Studying a Study and Testing a Test—How to Read the Medical Evidence, 4th ed. Philadelphia: Lippincott Williams & Wilkins, 2000.
2. Hennekens CH, Buring JE. Epidemiology in Medicine. Boston: Little, Brown and Company, 1987.

17. Answer: C

The SEM calculated from a single sample is determined by dividing the standard deviation by the square root of the sample size. For this case, then, 17 is divided by 6.93 to get an SEM of 2.45 (Answer C). The other answers simply represent various calculation errors: 0.35 (Answer A) is the standard deviation divided by the mean; 1.39 (Answer B) results from dividing the mean by the sample size; and 6.93 (Answer D) represents the result of dividing the sample size by the square root of the sample size.

1. Riegelman RK. Studying a Study and Testing a Test—How to Read the Medical Evidence, 4th ed. Philadelphia: Lippincott Williams & Wilkins, 2000.
2. Rosner B. Fundamentals of Biostatistics, 3rd ed. Boston: PWS-Kent Publishing, 1990.

18. Answer: A

For this type of comparison, the H_0 posits no difference between groups (Answer A). If the statistical comparison of the difference observed between groups is unlikely attributable to chance, then the H_0 is rejected, and a statistically significant difference can be concluded. Only Answer A reflects this definition. Answer B (that drug X and drug Y are not equally effective) is the definition for the alternative hypothesis in this case. Answer C and Answer D also could be the H_0 if the alternative hypothesis of interest were that drug X was more effective (Answer C) or that drug X was less effective (Answer D).

1. Gaddis ML, Gaddis GM. Introduction to biostatistics: part 3, sensitivity, specificity, predictive value, and hypothesis testing. Ann Emerg Med 1990;19:591–7.
2. Riegelman RK. Studying a Study and Testing a Test—How to Read the Medical Evidence, 4th ed. Philadelphia: Lippincott Williams & Wilkins, 2000.

19. Answer: C

Not all statistically significant results are clinically important. If a large enough sample is studied, statistically significant differences can be demonstrated for small changes. In this large study, a difference in hemoglobin A1C of 0.06% was found to be statistically significant. However, changes of this magnitude are seldom measured with such precision in clinical practice. Moreover, changes in this small range are not associated with clinically different outcomes. As such, Answer C best describes the appropriate interpretation of these results. The other answers mislabel the results either as not statistically significant (Answer A and Answer D) or as clinically significant (Answer A and Answer B).

1. Primer on statistical significance and p values. Eff Clin Pract 2001;4:183–4.
2. Froehlich GW. What is the chance that this study is clinically significant? A proposal for Q values. Eff Clin Pract 1999;2:234–9.

20. Answer: D

A 70% increased risk of miscarriage (Answer A) is incorrect since it is consistent with the upper bound of the CI, not the point estimate of the OR. The point estimate of the OR is 1.3. It can be said, given this value, that the risk is increased by 30% (Answer B). A 10% decreased risk of miscarriage with this drug (Answer C) might result if the lower bound of the CI were considered. However, in examining the 95% CI reported with the OR, it is seen that an OR of 1 cannot be excluded as representing the real relative odds for the groups (i.e., the CI range encompasses the value of 1). Therefore, the findings are not statistically significant (Answer D). In prospective trials using continuous variables, a CI that includes 0 would be interpreted as no difference.

1. Primer on 95% confidence intervals. Eff Clin Pract 2001;4:229–31.
2. Sackett DL, Straus SE, Richardson WS, Rosenberg W, Haynes RB. Evidence-Based Medicine: How to Practice and Teach EBM, 2nd ed. Edinburgh: Churchill Livingstone, 2000.

21. Answer: B

In trials that measure end points with a high degree of precision, that measure end points with a small amount of variation between groups, or that measure end points in a large number of patients, the ability to detect a statistically significant difference is often increased. Because not all statistically significant changes represent clinically important findings, results must always be viewed within a clinical context. Here, a difference of 20 calories during large energy expenditures is not clinically important (Answer B). Answer A is incorrect because decisions about the use of drugs or herbals should never be made only on the basis of a p value. Answer C is not consistent with the underlying premise of hypothesis testing because it assumes that conclusions can be based on a simple inspection of the observed differences. Answer D is incorrect because the results are statistically significant.

1. Primer on statistical significance and p values. Eff Clin Pract 2001;4:183–4.
2. Froehlich GW. What is the chance that this study is clinically significant? A proposal for Q values. Eff Clin Pract 1999;2:234–9.

CHOOSING THE APPROPRIATE STATISTICAL TEST

1. Answer: B

Answer B is correct because the data are categorical, and the Fisher exact test is used when the expected value of at least one cell of a 2 × 2 table is less than 5. Answer A (z-test) is incorrect because this statistical test is used when the data are normally distributed, the variance is known, and the data are continuous, not categorical, as in this case. Answer C (t-test) is incorrect because this statistical test is used when data are continuous and are relatively normally distributed and when the variance is unknown. Answer D (ANOVA) is incorrect because this test is a generalization of the t-test for more than two groups.

1. Rosner B. Fundamentals of Biostatistics, 6th ed. New York: Duxbury Press, 2005.
2. Strater R, Kurnik K, Heller C, Schobess R, Luigs P, Nowak-Gottl U. Aspirin versus low-dose low-molecular-weight heparin: antithrombotic therapy in pediatric ischemic stroke patients: a prospective follow-up study. Stroke 2001;32:2554–8.

2. Answer: C

Answer C is correct because the ANOVA test is a generalization of the t-test for more than two groups. This study compares four age groups. Answer A is incorrect because the Kruskal-Wallis test is a nonparametric analog of the ANOVA procedure, which is used for an ordinal dependent variable. Answer B (chi-square) is incorrect because height is measured on a continuous, not categorical, scale. Chi-square tests can only be used with categorical data. Answer D (paired t-test) is incorrect because this test is used for comparing data from just two groups when data are paired, as in a pre-post design. Neither of these conditions applies in this case.

1. Kleinbaum DG, Kupper LL, Muller KE, Nizam A. Applied Regression Analysis and Multivariable Methods, 3rd ed. New York: Duxbury Press, 1998.
2. Rosner B. Fundamentals of Biostatistics, 6th ed. New York: Duxbury Press, 2005.

3. Answer: C

Answer C (logistic regression) is correct. The outcome of interest was whether the patient had severe pain at his/her last observation. Answer A (linear regression) is incorrect because this technique is used when the outcome is measured on a continuous scale. Answer B (survival analysis) is incorrect because this approach is used when the outcome of interest is time to an event and when the analyst wants to account for censoring. Answer D (ordered logistic regression) is incorrect because this type of model is a generalization of logistic regression that allows the outcome to have more than two ordered categories.

1. Kleinbaum DG, Kupper LL, Muller KE, Nizam A. Applied Regression Analysis and Multivariable Methods, 3rd ed. New York: Duxbury Press, 1998.
2. Kleinbaum DG, Klein M, Pryor ER. Logistic Regression: A Self-Learning Text, 2nd ed. New York: Springer, 2005.
3. Clark TG, Bradburn MJ, Love SB, Altman DG. Survival analysis part I: basic concepts and first analyses. Br J Cancer 2003;89:232–8.

4. Answer: C

Answer C (the model explains 32% of the variation in the outcome) is correct because the r^2 is an estimate of the variability in the dependent variable that is explained by the independent variable(s). Answer A is incorrect because an r^2 of 0.57 would be needed for this statement to be true. In this case, $r = 0.57$, not r^2. Answer B (no conclusions can be drawn without knowing whether the coefficients for each covariate were statistically significant) is incorrect because consideration of statistical significance and accounting for variability in the dependent variable are different issues, tested against different null hypotheses. Answer D (the model explains 75% of the variation in the outcome) is incorrect because the square root of the correlation coefficient (r) is not used as a measure of variability in the dependent variable.

1. Kleinbaum DG, Kupper LL, Muller KE, Nizam A. Applied Regression Analysis and Multivariable Methods, 3rd ed. New York: Duxbury Press, 1998.
2. Hosmer DW Jr, Lemeshow S. Applied Logistic Regression, 2nd ed. New York: John Wiley & Sons, 2000.

5. Answer: A

Answer A (Kruskal-Wallis) is correct because this is the nonparametric analog to the ANOVA test, and the outcome is ordinal. Answer B (multiple ANOVA) is incorrect because the multiple ANOVA is used for normally distributed data when there are more than two groups and for multiple outcomes. Answer C is incorrect because the ANOVA is used for normally distributed data when there are more than two groups. Answer D is incorrect because the ANCOVA procedure is used when data are normally distributed and many independent variables exist, some of which are measured on a continuous scale.

1. Kleinbaum DG, Kupper LL, Muller KE, Nizam A. Applied Regression Analysis and Multivariable Methods, 3rd ed. New York: Duxbury Press, 1998.
2. Rosner B. Fundamentals of Biostatistics, 6th ed. New York: Duxbury Press, 2005.

6. Answer: C

Answer C (results from the post hoc analyses should be adjusted for multiple comparisons before being reported) is correct because multiple comparisons methods help prevent statistically significant findings from being found by chance. Answer A is incorrect because results from the post hoc analyses may be reported, but they should be identified as post hoc, exploratory results. Answer B (results from the post hoc analyses should be reported in the same way as the analyses planned before starting the study) is incorrect because results from the post hoc analyses should clearly indicate that those data were obtained from exploratory analyses. Answer D (results from the post hoc analyses should be reported only qualitatively) is incorrect because results from the post hoc analyses can be reported quantitatively.

1. Kleinbaum DG, Kupper LL, Muller KE, Nizam A. Applied Regression Analysis and Multivariable Methods, 3rd ed. New York: Duxbury Press, 1998.
2. Rosner B. Fundamentals of Biostatistics, 6th ed. New York: Duxbury Press, 2005.

7. Answer: A

Answer A (22 individuals) is correct because the outcome of interest is some harm, and it is calculated as the inverse of the absolute risk increase. In this case, the absolute risk increase is calculated by subtracting the outcome rate in the control group (3.7%) from that in the intervention group (8.2%), giving a difference of 4.5%. Expressed as a decimal, that gives 0.045, and the inverse, or number needed to harm, is 22.2. Answer B (0.2 individuals) is incorrect because the absolute risk increase needs to be expressed in decimal terms before the inverse is taken. Answer C (12 individuals) is incorrect because this estimate is the inverse of the outcome rate in the intervention group. Answer D (27 individuals) is incorrect because this intervention is the inverse of the outcome rate in the control group.

1. Haynes RB, Sackett DL, Guyatt GH, Tugwell P. Clinical Epidemiology: How to Do Clinical Practice Research, 3rd ed. Philadelphia: Lippincott Williams & Wilkins, 2006.
2. McAlister FA, Straus SE, Guyatt GH, Haynes RB. Users' guides to the medical literature: XX. Integrating research evidence with the care of the individual patient. JAMA 2000;283:2829–36.

8. Answer: B

Comparing nominal data is done using certain statistical tests developed for this purpose. Having or not having an exacerbation is a dichotomous type of outcome that meets the definition of nominal data. In this case, to compare two groups of nominal data, a chi-square test (Answer B) would be chosen. The McNemar test (Answer A) serves a role in statistical analysis similar to the chi-square, but it is used for data that are paired or matched (i.e., non-independent observations). A two-sample t-test (Answer C) is a parametric test used to compare two groups whose outcomes were measured on a continuous scale. The Mann-Whitney U-test is a nonparametric test, but it is used for comparisons using ordinal data.

1. Greenhalgh T. How to read a paper: statistics for the non-statistician. 1. Different types of data need different statistical tests. BMJ 1997;315:364–6.
2. Gaddis ML, Gaddis GM. Introduction to biostatistics: part 5, statistical inference techniques for hypothesis testing with nonparametric data. Ann Emerg Med 1990;19:1054–9.

9. Answer: D

Because post hoc data analysis can suffer from analytic shortcomings (e.g., lack of power or multiple comparisons), it typically should not be used to develop treatment strategies for groups of patients and is best regarded as hypothesis generating. In this study of a new antidepressant drug, the marginal significance of the p value (p=0.04), coupled with the post hoc data analysis, precludes conclusions regarding the efficacy of the drug in men versus women (Answer A). Designing a trial to test whether the drug has different efficacy in one sex or the other makes sense in this case (Answer D). Whether outcomes are considered primary or secondary is decided during the design phase of a trial and should not be altered on the basis of results obtained from data analysis at the end of the trial (Answer B and Answer C).

1. Bulpitt CJ. Subgroup analysis. Lancet 1988;2:31–4.
2. Oxman AD, Guyatt GH. A consumer's guide to subgroup analyses. Ann Intern Med 1992;116:78–84.

10. Answer: C

As do health care practitioners, patients need more than general information about the effects of drugs to make informed decisions. Statements such as "cardiovascular events increased" in women who took hormone replacement therapy do not provide sufficient detail about the likelihood of such a risk (Answer A). Although patients have intuitively understood results expressed as relative changes (Answer D), such figures alone do not convey any information about the underlying rate of events. Results expressed as risks or other ratios do not lend themselves to easy interpretation for much the same reason (Answer B). A result explained as a number needed to treat or other absolute change in the rate or incidence of an event is considered more useful in weighing risks and benefits (Answer C).

1. Steiner JF. Talking about treatment: the language of populations and the language of individuals. Ann Intern Med 1999;130:618–22.
2. Hux JE, Naylor CD. Communicating the benefits of chronic preventive therapy: does the format of efficacy data determine patients' acceptance of treatment? Med Decis Making 1995;15:152–7.
3. Chatterton HT. Efficacy, risk, and the determination of value—shared decision making in the age of information. J Fam Pract 1999;48:505–7.

11. Answer: A

Patients in this trial will be randomized to different treatment groups and not crossed over to the other group. Thus, the trial design measures the effect of the two drugs in each separate group of participants. With this knowledge, it can be concluded that these represent independent samples. Nausea and vomiting are graded using a ranking of symptoms; the data collected would be considered ordinal. Because the researchers are collecting data from two independent groups assessed on an ordinal scale, Table 3 in this chapter helps the reader deduce that an appropriate test is the nonparametric Wilcoxon rank sum test (Answer A). The Student t-test (Answer C) is the parametric equivalent of this test and is applied to continuous data. The Wilcoxon signed rank test (Answer B) and the paired Student t-test (Answer D) are both used for paired data. The former is used with ordinal data, and the latter is used with continuous data.

1. Gaddis ML, Gaddis GM. Introduction to biostatistics: part 5, statistical inference techniques for hypothesis testing with nonparametric data. Ann Emerg Med 1990;19:1054–9.
2. Riegelman RK. Studying a Study and Testing a Test—How to Read the Medical Evidence, 4th ed. Philadelphia: Lippincott Williams & Wilkins, 2000.

12. Answer: C

The results of this investigation into the effects of a drug on the incidence of stroke have been summarized in a Forest plot. Each dot and line pair on the figure in the question represent the mean difference compared with no use of a drug and 95% confidence interval (CI), respectively, for dissimilarities in stroke incidence in each of the individual trials included in this meta-analysis. The box at the bottom of the graph represents the real difference, as suggested by the results of the meta-analysis, together with a CI, again a line, for this estimate. Because the CI for this summary estimate of effect does not include the value for no difference (i.e., the value 1), the findings can be said to be statistically significant, which excludes Answer A and Answer D from consideration. Because the odds ratio (OR) associated with this drug's effect on stroke is greater than 1, it can be inferred that the risk of stroke is increased with the use of this drug (Answer C). Answer B would be correct only if a protective effect on stroke had been observed for this drug (i.e., an OR less than 1).

1. Lewis S, Clarke M. Forest plots: trying to see the wood and the trees. BMJ 2001;322:1479–80.
2. Etminan M, Levine M. Interpreting meta-analyses of pharmacologic interventions: the pitfalls and how to identify them. Pharmacotherapy 1999;19:741–5.

13. Answer: D

This question addresses concerns about presenting data as either relative changes or absolute changes. Because relative changes give no indication of the baseline frequency of an outcome, they should not be used as the sole basis for clinical or drug policy decisions; therefore, Answer A is incorrect. This representation of differences in hospitalization rates is the relative change that occurred between the two groups. It is the result of dividing the observed difference in admissions (5%) by the baseline rate (11%). Answer B represents the miscalculation of the relative change where the difference in admissions was divided by the rate of admissions in the group receiving the new drug.

Answer C and Answer D both come from attempts to calculate the number needed to treat from the absolute difference in events between the groups. The number needed to treat is the reciprocal of the absolute difference between groups. Answer C is a miscalculation of that value, resulting from a misplaced decimal when converting 5% to its decimal form (i.e., a conversion to 0.5 instead of 0.05). Answer D is the best answer and reports the correctly calculated number needed to treat of 20 patients to prevent one admission.

1. Chatellier G, Zapletal E, Lemaitre D, Menard J, Degoulet P. The number needed to treat: a clinically useful nomogram in its proper context. BMJ 1996;312:426–9.
2. Wiffen PJ, Moore RA. Demonstrating effectiveness—the concept of numbers-needed-to-treat. J Clin Pharm Ther 1996;21:23–7.
3. McQuay HJ, Moore RA. Using numerical results from systematic reviews in clinical practice. Ann Intern Med 1997;126:712–20.

14. Answer: D

This question describes an experiment that uses patients as their own controls to assess the effects of a new blood pressure drug. This is an efficient design for such comparisons, which permits a smaller sample size with less variability because patients are, in effect, matched to themselves to evaluate drug effects. Such a design, although it makes good sense in this case, must be analyzed with test statistics developed for paired or matched data. Although a paired t-test is designed to make such comparisons, the two-sample t-test is not. Because an inappropriate test was chosen to analyze the results of the trial, any conclusions are unreliable and potentially misleading (Answer D). Therefore, Answer A, Answer B, and Answer C are incorrect.

1. Gaddis ML, Gaddis GM. Introduction to biostatistics: part 5, statistical inference techniques for hypothesis testing with nonparametric data. Ann Emerg Med 1990;19:1054–9.

2. Greenhalgh T. How to read a paper: statistics for the non-statistician. 1. Different types of data need different statistical tests. BMJ 1997;315:364–6.

15. Answer: A

This question describes a correlation analysis for two variables. It asks how much of the change in forced expiratory volume in 1 second associated with the use of a certain chemotherapeutic drug can be explained by the dosage? To perform such an analysis, researchers must first determine the type of variables being used. Variables with different levels of measurement require different statistical tests to characterize any association between them. In this case, the variables of forced expiratory volume in 1 second and dosage are measured on continuous scales. The Pearson product moment coefficient (Answer A) is recommended for such a combination. The Spearman rank correlation (Answer C) would be used when assessing a proposed correlation between ordinal variables. An ANOVA and an ANCOVA (Answer B and Answer D) represent inferential statistical tests used when comparing more than three groups measured using continuous variables.

1. Gaddis ML, Gaddis GM. Introduction to biostatistics: part 6, correlation and regression. Ann Emerg Med 1990;19:1462–8.
2. Hurd PD. Research methodology: some statistical considerations. J Manag Care Pharm 1998;4:617–21.

16. Answer: B

The p value reported with the r for this association shows that the association between the dosage of this drug and forced expiratory volume in 1 second is statistically different from no correlation (i.e., $r=0$). In determining the amount of the change in forced expiratory volume in 1 second explained by changes in dosage, r^2 needs to be calculated. In this case, r^2 equals 0.21 and can be interpreted to mean that 21% of the variation in forced expiratory volume in 1 second is associated with changes in the dose of the drug (Answer B). The negative sign reported with r indicates that the relationship is an inverse one such that increasing dosages of drug are correlated with decreases in forced expiratory volume in 1 second. Therefore, Answer B is correct. Answer A (7%) is incorrect because it is the square root of r; Answer C (46%) is incorrect because it is simply the raw r value; and Answer D (92%) is incorrect because instead of squaring r, it multiplies r by 2.

1. BMJ Publishing Group. Statistics at Square One. Available at *http://bmj.bmjjournals.com/collections/statsbk/index.shtml*. Accessed January 13, 2010.
2. Gaddis ML, Gaddis GM. Introduction to biostatistics: part 6, correlation and regression. Ann Emerg Med 1990;19:1462–8.

17. Answer: B

The results reported for women older than 55 years represent a post hoc subgroup analysis. This type of data examination should be viewed as hypothesis generating, and differences observed as a result should be subject to further prospective investigation (Answer B). Answer A is incorrect because these types of analysis are more likely to produce type I errors (incorrectly concluding there is a difference when no true difference exists) than are examinations of the primary end point around which the study has been designed. Answer C and Answer D describe misinterpretations of the results with inappropriate extrapolations.

1. Koretz RL. The reading corner. IV. Subgroup analyses. Nutr Clin Pract 1998;13:230–4.
2. Oxman AD, Guyatt GH. A consumer's guide to subgroup analyses. Ann Intern Med 1992;116:78–84.

18. Answer: D

This question describes the results from a post hoc as-treated analysis. As with a subgroup analysis, such an approach to handling data that are not considered part of the original trial design does not yield reliable results. Moreover, an as-treated approach to analyzing the data has significant shortcomings even when planned beforehand. By switching the data for patients who were not at least 60% adherent to the placebo group, the initial randomization of this trial was corrupted. It is conceivable that the patients who were lacked adherence were somehow systematically different from those who did adhere. Were they more likely to suffer adverse effects? Were they sicker to begin with? Were they different in other ways? Obviously, the results of the as-treated analysis are susceptible to many biases. As such, they should serve as proposing a hypothesis for future prospective studies. Because the original intention-to-treat analysis found no difference between groups, Answer A, Answer B, and Answer C are incorrect. Answer D (wait and see) is the best approach on the basis of these results.

1. Sheiner LB, Rubin DB. Intention-to-treat analysis and the goals of clinical trials. Clin Pharmacol Ther 1995;57:6–15.
2. Gibaldi M, Sullivan S. Intention-to-treat analysis in randomized trials: who gets counted? J Clin Pharmacol 1997;37:667–72.

INTERPRETING RESULTS FROM CLINICAL TRIALS

1. Answer: C

A meta-analysis often is conducted with results of randomized, controlled trials, and the results may be reported using an OR. As with epidemiologic trial designs, CIs for ORs in meta-analyses can be interpreted as statistically significant only if they exclude the value of 1. In this systematic review, bisphosphonates decreased the use of radiation therapy in a statistically significant fashion. Their influence on spinal cord compression was not statistically significant (Answer C). Answer A, Answer B, and Answer D are incorrect because they represent interpretations of the CIs for the use of radiation therapy or spinal cord compression that are opposite of their true meanings.

1. Etminan M, Levine M. Interpreting meta-analyses of pharmacologic interventions: the pitfalls and how to identify them. Pharmacotherapy 1999;19:741–5.
2. Reith CH, Malone DC. Understanding the fundamental concepts for interpreting or conducting meta-analyses. Formulary 2001;36:594–6, 604–5, 609.

2. Answer: B

The results of meta-analyses wholly depend on the data that are included in the systematic review. In this meta-analysis, the trials represent a large range of sample sizes, which can make the results susceptible to the influence of one trial. In addition, so many trials are unlikely to have exactly the same types of participants or identical outcome measures. To help assess these challenges to the validity of any conclusions, sensitivity analysis and tests to detect heterogeneity between studies should be carried out (Answer B). A regression analysis (Answer A) is used to assess correlation or develop predictive models. A calculation of the hazard function for the total number of patients (Answer C) and a Cox regression analysis (Answer D) are used in the construction of survival curves and are not considered techniques used in conducting a meta-analysis.

1. Egger M, Smith GD, Phillips AN. Meta-analysis: principles and procedures. BMJ 1997;315:1533–7.
2. Eysenck HJ. Meta-analysis and its problems. BMJ 1994;309:789–92.

3. Answer: A

Literature evaluation is essential before using the results to make a decision that affects patient care. Study inclusion and exclusion criteria must be stated clearly and completely for a pharmacist to determine whether the results can be extrapolated to his or her patient population. It is possible that the patients included in the study were not similar to those seen in your hospital (Answer A). Type II errors occur when the authors state that no differences exist between the treatment groups (retain null hypothesis) when, in fact, there is a difference (Answer B). The credentials of the authors should always be reviewed to determine their competency to perform the study (Answer C). An evaluation of the adequacy to which blinding was used is necessary to determine whether investigator bias could have occurred regardless of whether an objective outcome was used (Answer D). Bias can be introduced during patient selection or with therapy evaluation, which could affect objective outcome measures. When evaluating the primary literature, it is important to evaluate critical aspects of an article, including the patient population, methods, statistics, and internal and external validity. A thorough evaluation will aid in determining whether the results can be applied to a specific patient.

1. Malone PM, Mosdell KW, Kier KL, Stanovich JE, eds. Drug Information: A Guide for Pharmacists, 2nd ed. New York: McGraw-Hill, 2001.
2. Altman DC, Schulz KF, Moher D, Egger M, Davidoff F, Gotzsche PZ, et al. The revised CONSORT statement for reporting randomized trials: explanation and elaboration. Ann Intern Med 2001;134:663–94.

4. Answer: B

Case reports or case series (Answer B) are the preferred type of literature to find detailed descriptions of adverse drug reactions. The case report often details the clinical presentation of the patient, circumstances leading to the reaction, and management techniques. Although randomized, controlled trials (Answer A) may contain information to support that a drug causes an adverse reaction, they do not provide sufficient details to guide the management of the reaction. Because Curemesis seldom causes Stevens-Johnson syndrome, it is unlikely that a case-control study (Answer D) has been conducted. The purpose of a meta-analysis (Answer C) is to combine primary literature statistically to derive an overall conclusion about a topic for which there are conflicting data or when the primary literature was underpowered. It is unlikely that a meta-analysis would be published on a rare adverse effect of a drug.

1. Malone PM, Mosdell KW, Kier KL, Stanovich JE, eds. Drug Information: A Guide for Pharmacists, 2nd ed. New York: McGraw-Hill, 2001.
2. Cuddy PG, Elenbaas RM, Elenbaas JK. Evaluating the medical literature. Part I: abstract, introduction, methods. Ann Emerg Med 1983;12:549–55.
3. Elenbaas JK, Cuddy PG, Elenbaas RM. Evaluating the medical literature. Part III: results and discussion. Ann Emerg Med 1983;12:679–86.
4. Riegelman RK. Studying a Study and Testing a Test: How to Read the Medical Evidence, 4th ed. Philadelphia: Lippincott Williams & Wilkins, 2000.

5. Answer: B

Because the study excluded patients with cardiovascular disease and the patients in your clinic have cardiovascular disease, your ability to apply the results of this trial to your patients is limited (Answer B). The inclusion criteria (Answer A) match well with the patients you see in your clinic, so they should not limit the applicability of the results. Although there are more male patients than female patients in this study, the distribution and mean age appear to be similar to the patients in your clinic (Answer C). That your patient population also includes patients who did not benefit from metformin will not limit the applicability of the results (Answer D).

1. Malone PM, Mosdell KW, Kier KL, Stanovich JE, eds. Drug Information: A Guide for Pharmacists, 2nd ed. New York: McGraw-Hill, 2001.
2. Riegelman RK. Studying a Study and Testing a Test: How to Read the Medical Evidence, 4th ed. Philadelphia: Lippincott Williams & Wilkins, 2000.
3. Guyatt G, Rennie D, eds. The Users' Guides to the Medical Literature: A Manual for Evidence-Based Clinical Practice. Chicago: American Medical Association, 2002.

6. Answer: A

This case describes an instance in which literature evaluation skills are critical. As the pharmacist on the Pharmacy and Therapeutics Committee, you should point out the limitations of retrospective trial designs (including the potential for recall bias [Answer A] and lower-quality data) and explain that the results of this study could be questionable and invalid. Drugs are not usually added to a formulary on the basis of a single study, even if a cost-savings is shown (Answer B). In addition, the hospital would have to examine its potential use of the drug to determine whether a cost-savings could be obtained in this hospital. Randomized, controlled trials, not retrospective studies, are the strongest for determining cause and effect (Answer C). Although comparative trials showed a small difference in heart rate favoring levalbuterol between the treatment groups, this difference does not warrant adding the drug to a formulary (Answer D).

1. Malone PM, Mosdell KW, Kier KL, Stanovich JE, eds. Drug Information: A Guide for Pharmacists, 2nd ed. New York: McGraw-Hill, 2001.
2. Altman DC, Schulz KF, Moher D, Egger M, Davidoff F, Gotzsche PZ, et al. The revised CONSORT statement for reporting randomized trials: explanation and elaboration. Ann Intern Med 2001;134:663–94.
3. Cuddy PG, Elenbaas RM, Elenbaas JK. Evaluating the medical literature. Part I: abstract, introduction, methods. Ann Emerg Med 1983;12:549–55.
4. Riegelman RK. Studying a Study and Testing a Test: How to Read the Medical Evidence, 4th ed. Philadelphia: Lippincott Williams & Wilkins, 2000.

7. Answer: A

In active-control trials, it is always important to check the doses of all drugs to ensure that they are the same as those used in clinical practice (Answer A). Because the two studies did not find a difference between the treatment groups, the reader should be concerned about type II error, not type I error (Answer B). Unless a power analysis was performed, there may have been insufficient patients to detect a difference between the groups when a difference exists. Although placebo-controlled studies give a true measure of efficacy, active-control trials also are valid because they compare new drugs with the current standard of care. In addition, the use of a placebo can be unethical when active treatment must be provided (e.g., epilepsy) (Answer C). The role of the study sponsor is always important to evaluate to identify bias (Answer D).

1. Malone PM, Mosdell KW, Kier KL, Stanovich JE, eds. Drug Information: A Guide for Pharmacists, 2nd ed. New York: McGraw-Hill, 2001.
2. Cuddy PG, Elenbaas RM, Elenbaas JK. Evaluating the medical literature. Part I: abstract, introduction, methods. Ann Emerg Med 1983;12:549–55.
3. Riegelman RK. Studying a Study and Testing a Test: How to Read the Medical Evidence, 4th ed. Philadelphia: Lippincott Williams & Wilkins, 2000.

8. Answer: D

Cross-sectional studies (Answer D) are surveys used to determine the prevalence of a disease at a specific time and are best suited for determining the prevalence of latex allergy in a population of pharmacy technicians. A randomized, controlled, clinical trial (Answer A) is best for determining the efficacy of a drug or treatment or for determining strong causal associations between exposures and disease (although this is often not feasible). A case-control study (Answer B) is a retrospective, observational trial that is used to identify an association between an exposure and the presence of a disease. Because the question specifically asks for the prevalence of latex allergy, a case-control study is not necessary. A cohort study (Answer C) is a prospective, observational study that monitors patients with an exposure to see whether a disease or event develops.

1. Malone PM, Mosdell KW, Kier KL, Stanovich JE, eds. Drug Information: A Guide for Pharmacists, 2nd ed. New York: McGraw-Hill, 2001.
2. Cuddy PG, Elenbaas RM, Elenbaas JK. Evaluating the medical literature. Part I: abstract, introduction, methods. Ann Emerg Med 1983;12:549–55.
3. Elenbaas JK, Cuddy PG, Elenbaas RM. Evaluating the medical literature. Part III: results and discussion. Ann Emerg Med 1983;12:679–86.
4. Riegelman RK. Studying a Study and Testing a Test: How to Read the Medical Evidence, 4th ed. Philadelphia: Lippincott Williams & Wilkins, 2000.

9. Answer: C

This question describes a common instance that occurs after the labeling of a drug is approved and the FDA requests further data on the safety of the drug. In this case, Superstatin is suspected of having an increased risk of liver toxicity, and a study needs to be conducted to determine whether such a risk exists. Because there were only a few previous reports, a retrospective trial cannot be conducted. A prospective cohort study (Answer C) is an observational trial used to monitor patients with an exposure to see whether a disease or event develops and is best suited for this question. A randomized, controlled, clinical trial (Answer A) is best for determining the absolute efficacy of a drug or treatment. A case-control study (Answer B) is a retrospective, observational trial used to identify an exposure with the presence of a disease. Cross-sectional studies (Answer D) are surveys that determine the prevalence of a disease at a specific point in time.

1. Malone PM, Mosdell KW, Kier KL, Stanovich JE, eds. Drug Information: A Guide for Pharmacists, 2nd ed. New York: McGraw-Hill, 2001.
2. Cuddy PG, Elenbaas RM, Elenbaas JK. Evaluating the medical literature. Part I: abstract, introduction, methods. Ann Emerg Med 1983;12:549–55.
3. Elenbaas JK, Cuddy PG, Elenbaas RM. Evaluating the medical literature. Part III: results and discussion. Ann Emerg Med 1983;12:679–86.
4. Riegelman RK. Studying a Study and Testing a Test: How to Read the Medical Evidence, 4th ed. Philadelphia: Lippincott Williams & Wilkins, 2000.

10. Answer: B

A case-control study (Answer B) is a retrospective, observational study that is conducted to identify an association of an exposure with the presence of a disease and is best suited to determine the risk of colon cancer with the use of low-dose aspirin. A retrospective trial would be best suited for this disease, which takes a long time to develop. A randomized, controlled, clinical trial is best for determining the efficacy of a drug or treatment. Because you wish to examine the risk of developing colon cancer while taking low-dose aspirin, a randomized, controlled trial (Answer A) would not be appropriate. A prospective cohort study (Answer C) is an observational study that is used to monitor patients with an exposure to see whether a disease or event develops. Because colon cancer takes a long time to develop, this study design would not be a preferred choice. Cross-sectional studies (Answer D) are surveys used to determine the prevalence of a disease at a specific time and therefore do not apply to this case.

1. Malone PM, Mosdell KW, Kier KL, Stanovich JE, eds. Drug Information: A Guide for Pharmacists, 2nd ed. New York: McGraw-Hill, 2001.
2. Cuddy PG, Elenbaas RM, Elenbaas JK. Evaluating the medical literature. Part I: abstract, introduction, methods. Ann Emerg Med 1983;12:549–55.
3. Elenbaas JK, Cuddy PG, Elenbaas RM. Evaluating the medical literature. Part III: results and discussion. Ann Emerg Med 1983;12:679–86.

4. Riegelman RK. Studying a Study and Testing a Test: How to Read the Medical Evidence, 4th ed. Philadelphia: Lippincott Williams & Wilkins, 2000.

11. Answer: A

A meta-analysis involves a thorough review of the published literature on a topic and includes a statistical analysis of the pooled results. Investigators of meta-analyses must be diligent in their search for articles to include in their analyses because articles with positive results are more likely to be published (publication bias) than those with negative results. Therefore, investigators must attempt to identify all research conducted on a topic through the use of many databases, conference proceedings, recognized experts, and reference lists (Answer A). A randomized, controlled trial is considered the gold standard for determining cause and effect. Meta-analyses can be helpful when previous studies could not enroll enough patients to detect a difference between the treatment groups; however, meta-analyses are considered more hypothesis generating than conclusive (Answer B). Because meta-analyses increase the sample size by pooling results from many trials, the risk of type II error (not finding a difference when one exists) is decreased (Answer C). The increased heterogeneity that occurs when many trials are pooled is a disadvantage of meta-analyses, possibly resulting in data that cannot be analyzed (Answer D).

1. Malone PM, Mosdell KW, Kier KL, Stanovich JE, eds. Drug Information: A Guide for Pharmacists, 2nd ed. New York: McGraw-Hill, 2001.
2. Egger M, Smith GD. Meta-analysis. Potentials and promise. BMJ 1997;315:1371–4.
3. Etminan M, Levine M. Interpreting meta-analyses of pharmacologic interventions: the pitfalls and how to identify them. Pharmacotherapy 1999;19:741–5.

12. Answer: A

Cigarette smoking is a confounding variable (Answer A) because it is related to the exposure (alcohol consumption) and may affect the outcome being studied (cancer). It is the most likely type of bias to occur in this kind of study. The external validity would not be affected (Answer B) by the inclusion of smokers in the study sample. The internal validity of this trial would be decreased because the investigators should have controlled for the number of smokers in each group or should have adjusted for it in their statistical analysis (Answer C). Prevalence bias occurs when time elapses between diagnosis or exposure and enrollment in the trial. This type of bias is not important in this trial because the patients were not given a diagnosis of cancer when they were enrolled in it (Answer D).

1. Guyatt G, Rennie D, eds. The Users' Guides to the Medical Literature: A Manual for Evidence-Based Clinical Practice. Chicago: American Medical Association, 2002.
2. Riegelman RK. Studying a Study and Testing a Test: How to Read the Medical Evidence, 4th ed. Philadelphia: Lippincott Williams & Wilkins, 2000.

13. Answer: D

Because the design of this study is retrospective, the biggest concern would be recall bias (Answer D). The patients will have to remember how much ephedra they took and how much weight they lost. The external validity (Answer A) of this trial is not a major concern because it is more important to analyze the methods of the study first. Blinding (Answer B) also is not an important concern with this type of study. Patients are not randomized (Answer C) in a case-control study.

1. Cuddy PG, Elenbaas RM, Elenbaas JK. Evaluating the medical literature. Part I: abstract, introduction, methods. Ann Emerg Med 1983;12:549–55.
2. Elenbaas JK, Cuddy PG, Elenbaas RM. Evaluating the medical literature. Part III: results and discussion. Ann Emerg Med 1983;12:679–86.
3. Riegelman RK. Studying a Study and Testing a Test: How to Read the Medical Evidence, 4th ed. Philadelphia: Lippincott Williams & Wilkins, 2000.

14. Answer: B

The most important threat to internal validity is that patients were selected into different groups (i.e., assigned to different drugs) systematically (Answer B). Perhaps one group of patients whose illness was thought to be more severe was assigned to drug A preferentially (or exclusively). Repeated testing (Answer A) does not play a large role in this case because the patients are self-reporting seizures. There is no identical test that patients can become familiar with over time. Experimental mortality is not a threat because investigators have included only patients with 60 days' worth of data. There will be no experimental mortality (Answer C) in this analysis. Instrumentation (Answer D) is not a threat because the seizure diaries in the question are assumed to be accurate.

1. Cook TD, Campbell DT. Quasi-experimentation. Boston: Houghton Mifflin, 1979.

2. Gliner JA, Morgan GA. Research Methods in Applied Settings: An Integrated Approach to Design and Analysis. Mahwah, N.J.: Lawrence Erlbaum Associates, 2000.

15. Answer: A

The valid conclusions made on the basis of this design are quite limited because of potential differences in the study populations. Therefore, all that can be said without further analysis is that patients taking drug A had fewer seizures during the study period (Answer A). It is inappropriate to make any broad inferences about the relative effectiveness of the drugs (Answer B and Answer C). Answer D is too limiting. Although the relative frequency of seizures can be calculated, care should be taken to avoid conclusions or inferences based on that calculation.

1. Andrews EB, Eaton S. Additional consideration in longitudinal database research. Value Health 2003;6:85–7.
2. Cook TD, Campbell DT. Quasi-experimentation. Boston: Houghton Mifflin, 1979.
3. Gliner JA, Morgan GA. Research Methods in Applied Settings: An Integrated Approach to Design and Analysis. Mahwah, N.J.: Lawrence Erlbaum Associates, 2000.
4. Rizzo JD, Powe NR. Methodological hurdles in conducting pharmacoeconomic analyses. Pharmacoeconomics 1999;15:339–55.
5. Else BA, Armstrong EP, Cox ER. Data sources for pharmacoeconomic and health services research. Am J Health Syst Pharm 1997;54:2601–8.

16. Answer: B

The BNP is a continuous variable (i.e., values are along a continuous scale with even and known distances between values); therefore, Answer A and Answer C are incorrect. The chi-square test is most appropriate for dichotomous variables (Answer D is incorrect). Answer B is the most appropriate statistical test with this type of variable and is the correct answer.

1. Norman GR, Streiner DL. PDQ Statistics, 3rd ed. Hamilton, Ontario, Canada: B.C. Decker, 2003.
2. Greenhalgh T. How to read a paper. Statistics for the non-statistician. I. Different types of data need different statistical tests. BMJ 1997;315:364–6.

17. Answer: C

Reporting continuous BNP values is more powerful than categorizing them, making Answer C correct. Dichotomous or categorical data are always inherently less powerful than continuous data because patients can only be in one state or another (e.g., alive or dead, not in heart failure or possibly in heart failure or in heart failure) and never in between (Answer A is incorrect). In general, even if it makes clinical sense to categorize data, it is a good idea to collect it in continuous form. Moreover, if a new study redefines the cutoff values (such as with fasting blood glucose in diabetes a few years ago), the data can always be re-categorized if they are continuous. Although surrogate outcome measures are usually continuous measures, this is not the reason to report actual BNP values (Answer B is incorrect). Even though it may be unlikely that a therapy would change BNP concentrations by 500 pg/mL, this is not the best reason to report actual BNP concentrations (Answer D is incorrect).

1. Nordness M. Epidemiology and Biostatistical Secrets. St. Louis: Mosby, 2005.
2. Norman GR, Streiner DL. PDQ Statistics, 3rd ed. Hamilton, Ontario, Canada: B.C. Decker, 2003.

18. Answer: A

The ARR is calculated as the difference between event rates in the control group and the treatment group. Recalling that these are percentages, the ARR can be expressed as 0.06 or 6% (Answer A is correct; Answer B, Answer C, and Answer D are incorrect).

1. Nordness M. Epidemiology and Biostatistical Secrets. St. Louis: Mosby, 2005.
2. Norman GR, Streiner DL. PDQ Statistics, 3rd ed. Hamilton, Ontario, Canada: B.C. Decker, 2003.

19. Answer: C

The RRR is 1 minus the relative risk, where relative risk is simply the ratio of event rates, treatment/control (in this case, 15%/21% = 0.71; and 1 - 0.71 = 0.29 or 29%), making Answer C correct and Answer A, Answer B, and Answer D incorrect.

1. Nordness M. Epidemiology and Biostatistical Secrets. St. Louis: Mosby, 2005.
2. Gordis L. Epidemiology, 4th ed. Philadelphia: Saunders Elsevier, 2008.

20. Answer: D

The NNT is calculated as 1/ARR (in this case, 1/0.06 = 17). More precisely, this would be stated as an NNT of 17 for 3 years to prevent one stroke (Answer D is correct; Answer A, Answer B, and Answer C are incorrect).

1. Nordness M. Epidemiology and Biostatistical Secrets. St. Louis: Mosby, 2005.
2. Gordis L. Epidemiology, 4th ed. Philadelphia: Saunders Elsevier, 2008.

21. Answer: D

The results are not statistically significant because the 95% CI crosses zero (Answer D is correct). Zero is the boundary at which p is greater than 0.05 (Answer B is incorrect). However, although this is a wide CI, not much information is provided (and it is subjective) (Answer A is incorrect). The finding is not statistically significant, although if the lower end of the 95% CI had been +1, this finding would have been correct. With the results shown, the authors have not excluded a reduction in stroke of up to 55% (not an increase in stroke—remember that these are RRRs, so a positive number is a reduction and a negative number is an increase) (Answer C is incorrect). So the correct statement is that the authors have not excluded a 1% increase in stroke.

1. Nordness M. Epidemiology and Biostatistical Secrets. St. Louis: Mosby, 2005.
2. Gordis L. Epidemiology, 4th ed. Philadelphia: Saunders Elsevier, 2008.

22. Answer: C

Power decreases with decreased sample size (Answer C is correct; Answer A, Answer B, and Answer D are incorrect). Usually, the greater the sample size, the higher the power. When a study is unable to get enough subjects to achieve the power the researchers originally anticipated, the chance of seeing a difference (power) drops. Decisions about the feasibility of the sample size with respect to recruitment speed, funding, and time should be made before the study is begun.

1. Friedman LM, Furberg CD, DeMets DL. Fundamentals of Clinical Trials, 3rd ed. New York: Springer, 1998.
2. Stolley PD, Strom BL. Sample size calculations in clinical pharmacology studies. Clin Pharmacol Ther 1986;39:489–90.
3. Campbell MJ, Julious SA, Altman DG. Estimating sample sizes for binary, ordered categorical, and continuous outcomes in two group comparisons. BMJ 1995;311:1145–8.

23. Answer: C

Repeat target vessel revascularization is a dichotomous outcome, so power is influenced by the efficacy of the treatment, event rate in the control group, and sample size. Thus, assuming the trial had an adequate sample size in the original protocol, two other places to introduce error would be in overestimating the efficacy of the new stent (underestimating would actually increase power, making Answer A incorrect) and overestimating the event rate in the control group (lower-risk patients would require a larger sample size; if the event rate were lower than initially assumed, then for the same sample size, the power would drop, making Answer C correct). Protocol violations and lack of blinding may adversely affect the quality of the study and may affect study power, but they are less likely (Answer B and Answer D are incorrect).

1. Friedman LM, Furberg CD, DeMets DL. Fundamentals of Clinical Trials, 3rd ed. New York: Springer, 1998.
2. Campbell MJ, Julious SA, Altman DG. Estimating sample sizes for binary, ordered categorical, and continuous outcomes in two group comparisons. BMJ 1995;311:1145–8.

24. Answer: A

The outcome (dependent) variable defines the most appropriate form of regression analysis (Answer A is correct). Continuous outcomes use various forms of linear regression, whereas dichotomous outcomes use various forms of logistic regression. The characteristics of the independent variables help refine the analysis to different forms of linear or logistic regression.

1. Nordness M. Epidemiology and Biostatistical Secrets. St. Louis: Mosby, 2005.
2. Norman GR, Streiner DL. PDQ Statistics, 3rd ed. Hamilton, Ontario, Canada: B.C. Decker, 2003.

25. Answer: C

Whether the independent variable is single or multiple is the best choice (Answer C is correct). Simple regression (either linear or logistic) considers only the impact of one factor (independent variable) on the outcome (dependent variable), whereas multiple independent variables require multiple regression techniques. As stated in question 24, the form of the outcome (continuous or dichotomous) defines whether linear or logistic regression techniques are used. Which clinical characteristics to enter in the model depend on the condition under study and the predefined analytic plan, but regardless of how this is determined, it is the number of characteristics that defines whether simple or multiple regression is used.

1. Nordness M. Epidemiology and Biostatistical Secrets. St. Louis: Mosby, 2005.
2. Norman GR, Streiner DL. PDQ Statistics, 3rd ed. Hamilton, Ontario, Canada: B.C. Decker, 2003.
3. Friedman LM, Furberg CD, DeMets DL. Fundamentals of Clinical Trials, 3rd ed. New York: Springer, 1998.

26. Answer: C

Adjust for baseline discrepancies is the best answer (Answer C is correct). In a randomized, controlled trial, if the univariate analysis of baseline variables such as patient demographics and comorbidities shows discrepancies in the two groups, then it is likely that such variables can affect the outcome (sometimes more than the treatment itself). Hence, to discover the true treatment effect, such variables should be adjusted for in the multiple regression model. Answer B is correct but only if Answer C is followed. Answer A would be true were it not a randomized, controlled trial, and Answer D is completely incorrect.

1. Nordness M. Epidemiology and Biostatistical Secrets. St. Louis: Mosby, 2005.
2. Norman GR, Streiner DL. PDQ Statistics, 3rd ed. Hamilton, Ontario, Canada: B.C. Decker, 2003.

27. Answer: A

Survival analysis is most likely to provide insight into the temporal course of treatment effects (Answer A is correct). By showing when events occur, a better understanding can be gained of the time course of treatment effects. Certainly, survival curves can also show the prognosis of patients with the disease under study, but this is not the main reason to use it (Answer B is incorrect). Although survival analyses almost always involve the tracking of "bad" events, there is no reason that any dichotomous event could not be tracked in this fashion (Answer C is incorrect). Survival analysis is also useful when patients are enrolled at different times and observed for different durations—allowing full use of the data (Answer D is incorrect).

1. Nordness M. Epidemiology and Biostatistical Secrets. St. Louis: Mosby, 2005.
2. Gordis L. Epidemiology, 4th ed. Philadelphia: Saunders Elsevier, 2008.
3. Campbell MJ, Julious SA, Altman DG. Estimating sample sizes for binary, ordered categorical, and continuous outcomes in two group comparisons. BMJ 1995;311:1145–8.

28. Answer: B

It would be best if the time until the loss to follow-up were still used (Answer B is correct). One advantage of survival analysis is that it uses data from every patient, even from those lost to follow-up. At the point at which they are lost, they are considered censored. This way, the time during which the patient was exposed to treatment (or control) and was at risk of an outcome event is still used, instead of throwing out that patient's data (Answer A is incorrect). Censoring does not assume the patient has had an outcome event (Answer C is incorrect), nor are any data imputed (Answer D is incorrect).

1. Nordness M. Epidemiology and Biostatistical Secrets. St. Louis: Mosby, 2005.
2. Friedman LM, Furberg CD, DeMets DL. Fundamentals of Clinical Trials, 3rd ed. New York: Springer, 1998.

29. Answer: D

A non-inferiority trial is best used when a placebo-controlled trial is not ethical or possible (Answer D is correct). An example would be for a new anticoagulant for deep venous thrombosis. Because it would be unethical to compare the new treatment against placebo, it would have to be compared with warfarin. If it seems unlikely that the new treatment is better than warfarin, then a non-inferiority design must be used (Answer B is incorrect). Such trials require higher sample sizes, so funding requirements would be higher (Answer C is incorrect). One condition of non-inferiority trials is that the comparison treatment must have already been established as efficacious; otherwise, the researcher might try to prove the equivalence of two useless therapies (Answer A is incorrect).

1. Kaul S, Diamond GA. Good enough: a primer on the analysis and interpretation of noninferiority trials. Ann Intern Med 2006;145:62–9.
2. Gomberg-Maitland M, Frison L, Halperin JL. Active-control clinical trials to establish equivalence or noninferiority: methodological and statistical concepts linked to quality. Am Heart J 2003;146:398–403.

30. Answer: C

Researchers set out to prove that treatment A is not worse than treatment B by a factor of Δ (Answer C is correct). Because it is impossible to prove that two treatments are exactly equivalent, new treatment A must be shown not to be worse than B by some factor, Δ. How Δ is chosen requires clinical judgment—essentially asking how much worse one treatment could be to say the treatments are "close enough."

1. Kaul S, Diamond GA. Good enough: a primer on the analysis and interpretation of noninferiority trials. Ann Intern Med 2006;145:62–9.
2. Gomberg-Maitland M, Frison L, Halperin JL. Active-control clinical trials to establish equivalence or noninferiority: methodological and statistical concepts linked to quality. Am Heart J 2003;146:398–403.

31. Answer: A

Systematic reviews reduce bias and are the highest level of evidence when high-quality randomized trials are included (Answer A is correct). Systematic reviews also follow a set protocol for searching, evaluating, extracting, and combining data. Narrative reviews, however, often mix opinions with evidence of variable quality, leading to bias. A good systematic review will often include a large number of studies and patients, combine data, and summarize the medical literature; however, these characteristics are not what reduce bias, thus leading to the high level of causal inference that systematic reviews provide (Answer B, Answer C, and Answer D are incorrect).

1. Oxman AD, Cook DJ, Guyatt GH, for the Evidence-Based Medicine Working Group. Users' guides to the medical literature. VI. How to use and overview. JAMA 1994;272:1367–71.
2. Greenhalgh T. How to read a paper: papers that summarise other papers (systematic reviews and meta-analyses. BMJ 1997;315:672–5.

32. Answer: A

The biggest pitfall is the quality of available trials (Answer A is correct). Simply put, garbage in equals garbage out. If the trials included are of poor methodological quality, the systematic review will be unreliable. Although analytic techniques continue to evolve, this is not a big pitfall (Answer B is incorrect). With a little practice, odds ratios are easily understood and applied to clinical practice (Answer C is incorrect). Although beyond the scope of this chapter, systematic review analytic techniques do not directly compare data between studies but rather within the same study, which allows a calculation of the "average" effect size (Answer D is incorrect).

1. Oxman AD, Cook DJ, Guyatt GH, for the Evidence-Based Medicine Working Group. Users' guides to the medical literature. VI. How to use and overview. JAMA 1994;272:1367–71.
2. Greenhalgh T. How to read a paper: papers that summarise other papers (systematic reviews and meta-analyses. BMJ 1997;315:672–5.

33. Answer: D

Surrogate outcomes may not represent the clinical outcome of interest (Answer D is correct). Because our understanding of pathophysiology is often incomplete, a surrogate outcome may not be truly part of the disease process and outcome of interest, or it may not be amenable to modification. Although surrogate outcomes are usually continuous variables, resulting in smaller sample size requirements (and less ability to detect rare adverse effects), these in themselves are not misleading (Answer A and Answer B are incorrect). Surrogate outcomes are usually based on pathophysiologic processes, and they can actually help enhance our understanding of the disease process (Answer C is incorrect).

1. Fleming TR, DeMets DL. Surrogate end points in clinical trials: are we being misled? Ann Intern Med 1996;125:605–13.
2. Valabhji J, Elkeles RS. Debate: are surrogate end-point studies worth the effort? Curr Control Trials Cardiovasc Med 2000;1:72–5.

34. Answer: D

Composite outcomes can reduce sample size requirement (Answer D is correct). Composite outcomes combine several different but related outcomes. This increases the event rate for this dichotomous outcome (i.e., how often the event of interest occurs), which reduces sample size requirements (Answer A is incorrect). They do not provide any more data on adverse effects (in fact, they may reduce the amount of data because of the smaller sample size) (Answer C is incorrect). Outcome events should not be added together because this causes "double or triple counting," whereby a patient experiencing many events would contribute excessively to the trial outcome (Answer B is incorrect). Rather, only the first or most important event, if there are several events, is usually counted.

1. Montori VM, Permanyer-Miralda G, Ferreira-Gonzalez I, Busse JW, Pacheco-Huergo V, Bryant D, et al. Validity of composite endpoints in clinical trials. BMJ 2005;330:594–6.

2. Kip KE, Hollabaugh K, Marroquin O, Williams DO. The problem with composite end points in cardiovascular studies: the story of major adverse cardiac events and percutaneous coronary intervention. J Am Coll Cardiol 2008;51:701–7.

35. Answer: B

Composite outcomes assume that all outcomes are equally important (Answer B is correct). By combining different events such as hospitalization, surgery, and death, the assumption is made that all events are equally clinically important. This may not always be the best assumption. To be fair, only the first (or most important, if that is decided) event in a cluster of events should be counted; to do otherwise is to count many events for the same patient and unduly contribute to the event rate (Answer A is incorrect). One reason to choose a composite outcome is because of the reduced sample size; this means less power to determine treatment effects on individual clinical outcomes (Answer C is incorrect). As with all clinical outcome trials, all outcomes must be classifiable or adjudicated (Answer D is incorrect); this is often done by a blinded outcomes assessment committee.

1. Montori VM, Permanyer-Miralda G, Ferreira-Gonzalez I, Busse JW, Pacheco-Huergo V, Bryant D, et al. Validity of composite endpoints in clinical trials. BMJ 2005;330:594–6.
2. Kip KE, Hollabaugh K, Marroquin O, Williams DO. The problem with composite end points in cardiovascular studies: the story of major adverse cardiac events and percutaneous coronary intervention. J Am Coll Cardiol 2008;51:701–7.

PHARMACOEPIDEMIOLOGY

1. Answer: D

Answer D is correct because the incidence rate is a measure of the number of study subjects who have the outcome of interest divided by the total time contributed by the individuals observed. Answer A is incorrect because the risk is the number of individuals who experience an outcome during a period divided by the number of individuals observed for that period. Answer B is incorrect because the hazard ratio is the instantaneous probability of experiencing the outcome assuming the person is at risk of the event up to that point. Answer C is incorrect because prevalence is a measure of outcome status that indicates how many outcomes of interest have occurred.

1. Rothman KJ. Measuring disease occurrence and causal effects. In: Epidemiology: An Introduction. New York: Oxford University Press, 2002:29.
2. Rothman KJ, Greenland S. Measures of disease frequency. In: Rothman KJ, Greenland S, eds. Modern Epidemiology, 2nd ed. Philadelphia: Lippincott-Raven Publishers, 1998:29–46.

2. Answer: B

Answer B is correct because prevalence is an estimate of the number of individuals who meet some specified criteria divided by the number of individuals in the population. Answer A is incorrect because period prevalence is a hybrid of incidence and prevalence. There is no reliable measure of incidence. Answer C is incorrect because incidence requires a reliable estimate of the number of new events divided by the total population at risk. Neither piece of information is here. Answer D is incorrect because incidence density requires not only the number of new cases, but also the total person-time of observation.

1. Rothman KJ. Measuring disease occurrence and causal effects. In: Epidemiology: An Introduction. New York: Oxford University Press, 2002:29.
2. Rothman KJ, Greenland S. Measures of disease frequency. In: Rothman KJ, Greenland S, eds. Modern Epidemiology, 2nd ed. Philadelphia: Lippincott-Raven Publishers, 1998:29–46.

3. Answer: B

Answer B is correct because the absolute risk reduction is likely to be the clearest measure to allow employers to see how avoiding the drug would affect the number of events experienced. Answer A is incorrect because relative risk reduction may be difficult to interpret. Answer C is incorrect because incidence is of use, but it may be difficult to interpret for employers. Answer D is incorrect because prevalence estimates are important, but they may be difficult for employers to interpret.

1. Rothman KJ. Measuring disease occurrence and causal effects. In: Epidemiology: An Introduction. New York: Oxford University Press, 2002:29.
2. Rothman KJ, Greenland S. Measures of disease frequency. In: Rothman KJ, Greenland S, eds. Modern Epidemiology, 2nd ed. Philadelphia: Lippincott-Raven Publishers, 1998:29–46.

4. Answer: C

Answer C is correct because cohort studies observe the experiences of a group of people who share a common experience or condition. Because the event of interest has already occurred, the investigators follow the cohort backward through time to determine its exposure status. Answer A is incorrect because a nested case-control study would have a clearly defined cohort and would include selection of control subjects. Answer B is incorrect because, although this example is a cohort study, to be prospective, observation would have to begin before the outcome of interest occurs. Answer D is incorrect because this experiment is neither a case-control study nor prospective.

1. Rothman KJ, Greenland S. Cohort Studies. In: Rothman KJ, Greenland S, eds. Modern Epidemiology, 2nd ed. Philadelphia: Lippincott-Raven Publishers, 1998:79–91.
2. Rothman KJ. Types of epidemiologic study. In: Epidemiology: An Introduction. New York: Oxford University Press, 2002:57–93.

5. Answer: B

Answer B, a case-control study, is correct because we have a relatively common exposure and a rare outcome. Answer A is incorrect because it would not be possible to conduct a randomized, controlled study in this setting, given the published evidence of serious risk. Answer C is incorrect because cohort studies are less efficient than case-control studies to study rare outcomes. It would likely be time-consuming and expensive to conduct a cohort study in this case. Answer D is incorrect because a case-cohort study requires that each person receive all treatments, allowing time for the drug effect to wear off between exposure periods. Given the evidence of serious risk, it would not be possible to do this type of study.

1. Rothman KJ, Greenland S. Cohort studies. In: Rothman KJ, Greenland S, eds. Modern Epidemiology, 2nd ed. Philadelphia: Lippincott-Raven Publishers, 1998:79–91.
2. Rothman KJ. Types of epidemiologic study. In: Epidemiology: An Introduction. New York: Oxford University Press, 2002:57–93.
3. Rothman KJ, Greenland S. Case-control studies. In: Rothman KJ, Greenland S, eds. Modern Epidemiology, 2nd ed. Philadelphia: Lippincott-Raven Publishers, 1998:93–114.

6. Answer: A

Answer A is correct because many databases used for retrospective analyses include data for a large number of individuals collected over years of observation. Answer B is incorrect because databases are not inherently unbiased. Answer C is incorrect because results from well-designed database analyses are comparable with those generated from well-designed, randomized, controlled studies. Answer D is incorrect because databases may lend themselves to imputation when data are missing, but the process is not necessarily easy, nor are these estimates automatically valid.

1. Motheral B, Brooks J, Clark MA, Crown WH, Davey P, Hutchins D, et al. A checklist for retrospective database studies—report of the ISPOR Task Force on Retrospective Databases. Value in Health 2003;6:90–97.
2. Strom BL, ed. Pharmacoepidemiology, 4th ed. Hoboken, N.J.: John Wiley & Sons, 2005.
3. International Society of Pharmacoepidemiology. Guidelines for good pharmacoepidemiology practices (GPP). Pharmacoepidemiol Drug Saf 2005;14:589–95.

7. Answer: B

Answer B is correct because people who are insured may differ from individuals who are uninsured in ways that make direct comparisons difficult. For example, individuals with health insurance have relatively easy access to health care, whereas people who are uninsured generally have more restricted access. People with employer-based health insurance are either employed or part of a household in which someone has a job that offers insurance as a benefit. These jobs may also require some level of training or education, which uninsured individuals may not have. Answer A is incorrect because confounding may be present, but it is not the main limitation. Answer C is incorrect because selection bias may interfere with interpretation of the findings, but is not the main limitation. Answer D is incorrect because information bias may interfere with interpretation of the findings, but it is not the main limitation.

1. Rothman KJ, Greenland S. Precision and validity in epidemiologic studies. In: Rothman KJ, Greenland S, eds. Modern Epidemiology, 2nd ed. Philadelphia: Lippincott-Raven Publishers, 1998:125–33.
2. Rothman KJ. Biases in study design. In: Epidemiology: An Introduction. New York: Oxford University Press, 2002:94–112.

8. Answer: B

Answer B is correct because the study population and analyses can be compared with similar groups. If the experimental groups are similar, generalizability is supported. Answer A is incorrect because the study population and analyses can be stratified, but that will not address the generalizability issue. Answer C is incorrect because restriction will not address generalizability. Answer D is incorrect because this type of restriction does not address the generalizability issue.

1. Rothman KJ, Greenland S. Precision and validity in epidemiologic studies. In: Rothman KJ, Greenland S, eds. Modern Epidemiology, 2nd ed. Philadelphia: Lippincott-Raven Publishers, 1998:125–33.
2. Rothman KJ. Biases in study design. In: Epidemiology: An Introduction. New York: Oxford University Press, 2002:94–112.

9. Answer: B

Answer B is correct because the number of cases and total at-risk population may be incompletely observed. Answer A is incorrect because well-designed population-based cohort studies may, but do not necessarily, provide valid and reliable data. Answer C is incorrect because bias may be present even though birth defects are generally uncommon, and the researchers had access to information from several sources. Answer D is incorrect because although confounding is likely present, it is not the most significant potential bias of these estimates.

1. Rothman KJ, Greenland S. Precision and validity in epidemiologic studies. In: Rothman KJ, Greenland S, eds. Modern Epidemiology, 2nd ed. Philadelphia: Lippincott-Raven Publishers, 1998:125–33.
2. Rothman KJ. Biases in study design. In: Epidemiology: An Introduction. New York: Oxford University Press, 2002:94–112.

10. Answer: B

Answer B is correct because adverse drug event reporting is often incomplete. Although this data set may identify all the individuals who received the drug of interest, it is likely that not all adverse drug events that occurred during the study period will be identified, resulting in misclassification of the outcome, which is a type of information bias. Answer A is incorrect because selection bias exists when study participation is influenced by factors related to how individuals were selected to participate in the study.

All individuals admitted to this medical center are included in this data set. Answer C is incorrect because confounding exists when a factor is related to exposure and outcome, is not in the causal pathway, and differs between groups. Answer D is incorrect because the researcher is not trying to apply generalizations about the population to individuals in this sample.

1. Rothman KJ, Greenland S. Precision and validity in epidemiologic studies. In: Rothman KJ, Greenland S, eds. Modern Epidemiology, 2nd ed. Philadelphia: Lippincott-Raven Publishers, 1998:125–33.
2. Kimmel SE, Sekeres MA, Berlin JA, Goldberg LR, Strom BL. Adverse events after protamine administration in patients undergoing cardiopulmonary bypass: risks and predictors of under-reporting. J Clin Epidemiol 1998;51:1–10.

11. Answer: C

Answer C is correct because confounding by indication exists when there are differences in prognosis between individuals given different therapies. In this case, if individuals who use the drugs in question are at increased risk of kidney dysfunction, confounding by indication may exist. Answer A is incorrect because immortal time bias occurs when individuals who are not at risk of the event of interest contribute time to the denominator. Because these individuals cannot experience the event, the observation time increases, but the number of observed events does not. The intervention being studied will look safer than it may be. Answer B is incorrect because diagnostic bias refers to a type of information bias. Answer D is incorrect because information bias refers to error in what or how data are collected on study participants.

1. Rothman KJ. Epidemiology: An Introduction. New York: Oxford University Press, 2002:206.
2. Signorello LB, McLaughlin JK, Lipworth L, Friis S, Sørensen HT, Blot WJ. Confounding by indication in epidemiologic studies of commonly used analgesics. Am J Ther 2002;9:199–205.

12. Answer: D

Answer D is correct because selection bias results from procedures used to select subjects and factors that contribute to study participation. Answer A is incorrect because confounding occurs when there is a factor related to the exposure and the outcome that is differentially distributed between experimental groups. Answer B is incorrect because information bias occurs when data are erroneous. Answer C is incorrect because immortal time bias occurs when individuals not at risk of the event are allowed to contribute observation time.

1. Rothman KJ. Epidemiology: An Introduction. New York: Oxford University Press, 2002:96–8.
2. Begg C, Cho M, Eastwood S, Horton R, Moher D, Olkin I, et al. Improving the quality of reporting of randomized controlled studies. The CONSORT statement. JAMA 1996;276:637–9.

13. Answer: B

Answer B is correct because this factor determines whether the misclassification is differential or non-differential. Answer A is incorrect because although the degree of misclassification is of concern, it does not automatically invalidate the results. Answer C is incorrect because the misclassified data do not differ by outcome. Answer D is incorrect because this information is useful, but not the most important.

1. Rothman KJ, Greenland S. Precision and validity in epidemiologic studies. In: Rothman KJ, Greenland S, eds. Modern Epidemiology, 2nd ed. Philadelphia: Lippincott-Raven Publishers, 1998:125–33.
2. Rothman KJ. Biases in study design. In: Epidemiology: An Introduction. New York: Oxford University Press, 2002:94–112.

14. Answer: C

Answer C is correct because the data are similarly prone to error for exposed and unexposed individuals. Answer A is incorrect because we cannot tell whether this is true from the data we have. Answer B is incorrect because the extent of misclassification is similar between individuals who were exposed and those who were not exposed. Answer D is incorrect because this is not the definition of differential misclassification.

1. Rothman KJ, Greenland S. Precision and validity in epidemiologic studies. In: Rothman KJ, Greenland S, eds. Modern Epidemiology, 2nd ed. Philadelphia: Lippincott-Raven Publishers, 1998:125–33.
2. Rothman KJ. Biases in study design. In: Epidemiology: An Introduction. New York: Oxford University Press, 2002:94–112.

15. Answer: D

Answer D is correct because this is the likely consequence of non-differential misclassification. Answer A is incorrect because it is differential misclassification. Answer B is incorrect because non-differential misclassification generally produces bias toward the null. Answer C is incorrect because this phenomenon is a possible effect of differential misclassification.

1. Rothman KJ, Greenland S. Precision and validity in epidemiologic studies. In: Rothman KJ, Greenland S, eds. Modern Epidemiology, 2nd ed. Philadelphia: Lippincott-Raven Publishers, 1998:125–33.
2. Rothman KJ. Biases in study design. In: Epidemiology: An Introduction. New York: Oxford University Press, 2002:94–112.

16. Answer: D

Answer D is correct because this action may help alleviate the misclassification by providing more data than either source contains individually. Answer A is incorrect because matching is a possible approach to control confounding, a phenomenon not described by this question. Answer B is incorrect because restriction is a possible approach to control confounding. Answer C is incorrect because misclassification is an important type of bias and should be addressed.

1. Rothman KJ, Greenland S. Precision and validity in epidemiologic studies. In: Rothman KJ, Greenland S, eds. Modern Epidemiology, 2nd ed. Philadelphia: Lippincott-Raven Publishers, 1998:125–33.
2. Rothman KJ, Greenland S. Matching. In: Rothman KJ, Greenland S, eds. Modern Epidemiology, 2nd ed. Philadelphia: Lippincott-Raven Publishers, 1998:147–61.

17. Answer: B

Answer B is correct because additional testing to rule out other possible causes of this outcome will help reduce misclassification of outcome. Answer A is incorrect because regardless of whether the misclassification was differential, it needs to be explored. Answer C is incorrect because randomization is not a likely option. Answer D is incorrect because collecting more information from patients would be helpful for misclassification of exposure, not of the outcome.

1. Rothman KJ, Greenland S. Precision and validity in epidemiologic studies. In: Rothman KJ, Greenland S, eds. Modern Epidemiology, 2nd ed. Philadelphia: Lippincott-Raven Publishers, 1998:125–33.
2. Friedman LM, Furberg CD, DeMets DL. Fundamentals of Clinical Trials, 3rd ed. New York: Springer, 1998.

18. Answer: A

Answer A (confounding) is correct because imbalances between experimental groups are a necessary condition for confounding. Answer B is incorrect because immortal time bias is caused by including individuals who are not at risk of the event in the incidence rate denominator. Answer C is incorrect because selection bias pertains to study participation. Answer D is incorrect because information bias refers to errors in what is known about the study participants.

1. Rothman KJ, Greenland S. Precision and validity in epidemiologic studies. In: Rothman KJ, Greenland S, eds. Modern Epidemiology, 2nd ed. Philadelphia: Lippincott-Raven Publishers, 1998:125–33.
2. Suissa S. Effectiveness of inhaled corticosteroids in chronic obstructive pulmonary disease. Immortal time bias in observational studies. Am J Resp Crit Care Med 2003;168:49–53.

19. Answer: B

Answer B is correct because randomization helps prevent confounding and, thus, channeling bias, by ensuring similar groups. Answer A is incorrect because we cannot tell whether the imbalanced factors are in the causal pathway. Answer C is incorrect because the presence of imbalanced factors related to the exposure is a condition for confounding. Answer D is incorrect because the presence of imbalanced factors related to the outcome is a condition for confounding.

1. Glynn RJ, Knight EL, Levin R, Avorn J. Paradoxical relations of drug treatment with mortality in older persons. Epidemiology 2001;12:682–9.
2. Petri H, Urquhart J. Channeling bias in the interpretation of drug effects. Stat Med 1991;10:577–81.

20. Answer: D

Answer D is correct because the exposure is binary (yes/no), and the estimate is closer to the null than other published results. Answer A is incorrect because selection bias refers to distortions of effect caused by procedures used to select study subjects and by factors that influence study participation. Answer B is incorrect because channeling bias refers to distortions that occur when a drug is channeled to (or away from) a specific group of people, as when newer drugs are reserved for individuals whose disease is more severe. Answer C is incorrect because confounding requires some imbalance of a factor related to the outcome between study groups. Confounding may be present, but there is not enough information given to know for sure.

1. Rothman KJ, Greenland S. Precision and validity in epidemiologic studies. In: Rothman KJ, Greenland S, eds. Modern Epidemiology, 2nd ed. Philadelphia: Lippincott-Raven Publishers, 1998:125–33.
2. Rothman KJ. Biases in study design. In: Epidemiology: An Introduction. New York: Oxford University Press, 2002:94–112.

21. Answer: B

Answer B is correct because case-control studies are well suited to studying rare events and are generally more efficient than cohort studies. In this case, the outcome of interest is rare, so a large cohort would need to be observed for a long time. In addition, collecting information on all non-cases in the source population would require substantial resources. Answer A is incorrect because cohort studies are generally less efficient than case-control studies for rare events. Answer C is incorrect because equipoise about this question no longer exists (or is at least in doubt). As a result, a randomized, controlled trial is not the best choice of study design for this question. Answer D is incorrect because the term *prospective* (or *retrospective*) refers only to the timing of the information and events of the study.

1. Koepsell TD, Weiss NS. Epidemiologic Methods. Studying the Occurrence of Illness. New York: Oxford University Press, 2003:107.
2. Rothman KJ. Epidemiology: An Introduction. New York: Oxford University Press, 2002:73–5.

22. Answer: A

Answer A is correct because confounding is caused by factors related to the exposure and the outcome. A table that shows the distribution of clinical and demographic factors between study groups is designed to show possible imbalances in factors that may be confounders. Answer B is incorrect because selection bias refers to the methods used to choose study subjects and factors that influence study participation. Answer C is incorrect because information bias occurs when the information collected from or about study subjects is wrong. Answer D is incorrect because effect modification refers to instances in which the effect of a factor on the outcome depends on the level of other factors. For example, if the effect of age on mortality depends on whether the individual is male or female, effect modification is present.

1. Rothman KJ. Epidemiology: An Introduction. New York: Oxford University Press, 2002:94–111.
2. Koepsell TD, Weiss NS. Epidemiologic Methods. Studying the Occurrence of Illness. New York: Oxford University Press, 2003:247–80.

23. Answer: B

Answer B is correct because a Consolidated Standards of Reporting Studies–style flowchart provides information about the number of study participants at each stage of the study. Answer A is incorrect because immortal time bias refers to the error introduced when person-time for individuals not at risk of the event of interest is included in the denominator. Answer C is incorrect because information bias occurs when data collected from or about study participants are wrong. Answer D is incorrect because confounding is caused by the differential distribution of factors related to the outcome that are not in the causal pathway.

1. Rothman KJ. Epidemiology: An Introduction. New York: Oxford University Press, 2002:94–111.
2. Koepsell TD, Weiss NS. Epidemiologic Methods. Studying the Occurrence of Illness. New York: Oxford University Press, 2003:247–80.

24. Answer: B

Answer B is correct because the rate ratio, calculated as incidence in exposed individuals divided by incidence in unexposed individuals, is 10. This estimate is interpreted as the incidence of the outcome being 10 times higher in exposed individuals than in unexposed individuals. Answer A is incorrect; it would be the interpretation if the rate ratio were calculated as incidence in unexposed individuals divided by incidence in exposed individuals. Answer C is incorrect because this interpretation reflects the incidence in the general population divided by the incidence in the unexposed group. Answer D is incorrect because this interpretation reflects the incidence in unexposed individuals divided by the incidence in the general population.

1. Rothman KJ. Epidemiology: An Introduction. New York: Oxford University Press, 2002:52.
2. Koepsell TD, Weiss NS. Epidemiologic Methods. Studying the Occurrence of Illness. New York: Oxford University Press, 2003:198–9.

25. Answer: A

Answer A is correct because the attributable fraction is calculated by subtracting the incidence in unexposed individuals from that in exposed individuals and dividing that result by the incidence in exposed individuals. The result is 0.9, indicating that the event rate in the general population is 10% of that in exposed individuals. Answer B is incorrect because this interpretation reflects the ratio of the sum of the incidence in the general population and the incidence in unexposed individuals divided by the incidence in exposed individuals. Answer C is incorrect because the attributable fraction is not calculated as the difference between the incidence in the general population and the incidence in unexposed individuals divided by the incidence in unexposed individuals. Answer D is incorrect because the attributable fraction is not calculated as the difference between the incidence in exposed individuals and that in unexposed individuals divided by the incidence in unexposed individuals.

1. Rothman KJ. Epidemiology: An Introduction. New York: Oxford University Press, 2002:52.
2. Koepsell TD, Weiss NS. Epidemiologic Methods. Studying the Occurrence of Illness. New York: Oxford University Press, 2003:198–9.

REVIEWERS

The editor and the American College of Clinical Pharmacy would like to thank the following individuals for their careful review.

G. Robert DeYoung, Pharm.D., BCPS
Saint Mary's Health Care and Advantage Health Physician Network
Grand Rapids, Michigan

Greg Stoddard, MBA, MPH, Ph.D.
University of Utah
School of Medicine
Salt Lake City, Utah

The editor and the American College of Clinical Pharmacy would like to express their appreciation to the authors of the original previously published material that has been adapted here into this text.

G. Robert DeYoung, Pharm.D., BCPS
Saint Mary's Health Care and Advantage Health Physician Network
Grand Rapids, Michigan

Sipi Garg, M.Sc.
University of Alberta
Edmonton, Alberta, Canada

Scott A. Strassels, Pharm.D., Ph.D., BCPS
University of Texas at Austin
Austin, Texas

Ross T. Tsuyuki, Pharm.D., MSc., FCSHP
University of Alberta
Edmonton, Alberta, Canada

James P. Wilson, Pharm.D., Ph.D.
University of Texas at Austin
Austin, Texas

DISCLOSURE OF POTENTIAL CONFLICTS OF INTEREST

Grants: Robert DiCenzo
(Merck; National Institute on Drug Abuse)

Honoraria: G. Robert DeYoung
(American College of Clinical Pharmacy [speaker])